Advances in Computer Vision and Pattern Recognition

Titles in this series now included in the Thomson Reuters Book Citation Index!

Advances in Computer Vision and Pattern Recognition is a series of books which brings together current developments in all areas of this multi-disciplinary topic. It covers both theoretical and applied aspects of pattern recognition, and provides texts for students and senior researchers in areas including, but not limited to:

- Computer vision
- Biological vision
- Signal processing
- Image processing and analysis
- Video processing and analysis
- Document analysis
- Character recognition
- Speech analysis and recognition
- Biometrics
- Multimedia
- Virtual reality
- Neural networks
- Machine vision and learning
- Robotics
- Industrial control and automation
- Genetic algorithms and evolutionary computing
- Intelligent forecasting

More information about this series at https://link.springer.com/bookseries/4205

Qi Wu · Peng Wang · Xin Wang · Xiaodong He ·
Wenwu Zhu

Visual Question Answering

From Theory to Application

 Springer

Qi Wu (ID)
School of Computer Science
University of Adelaide
Adelaide, SA, Australia

Peng Wang (ID)
Department of Computer Science
Northwestern Polytechnical University
Xi'an, Shaanxi, China

Xin Wang (ID)
Tsinghua University
Beijing, China

Xiaodong He
AI Lab
JD.COM
Beijing, China

Wenwu Zhu
Department of Computer Science
Tsinghua University
Haidian, China

ISSN 2191-6586   ISSN 2191-6594 (electronic)
Advances in Computer Vision and Pattern Recognition
ISBN 978-981-19-0966-5   ISBN 978-981-19-0964-1 (eBook)
https://doi.org/10.1007/978-981-19-0964-1

This Springer imprint is published by the registered company Springer Nature Singapore Pte Ltd.
The registered company address is: 152 Beach Road, #21-01/04 Gateway East, Singapore 189721,
Singapore

For the people who have lost their lives during the COVID-19 pandemic, including my father, Zhiwei Wu, who always supported me in life.

— Qi Wu

Preface

Visual question answering (VQA) is a fundamental task in vision-and-language research and has attracted considerable attention from computer vision (CV), natural language processing (NLP) and other diverse artificial intelligence communities. VQA connects CV and NLP, thereby stimulating research and expanding the limits of both fields. In the most common form of VQA, the computer is presented with an image and a textual question regarding the image. Subsequently, the computer must determine the correct answer and present it in a few words or a short phrase. Variants include binary (yes/no) and multiple-choice settings, in which candidate answers are proposed.

A key distinction between VQA and other tasks in computer vision is that the question to be answered is not determined until run time. In traditional problems such as segmentation or object detection, the question to be answered by an algorithm is predetermined, and only the input image changes. In contrast, in VQA, the form of the question and set of operations required to answer it are unknown. This task is associated with the challenge of general image understanding. In particular, VQA is related to the task of textual question answering, in which the answer must be sought in a specific textual narrative (i.e., reading comprehension) or large knowledge bases (i.e., information retrieval). Textual QA has been studied for a long time by the NLP community, and VQA represents its extension to additional visual supporting information. Notably, this extension is accompanied by a significant challenge, as images have a larger number of dimensions and more noise than pure text. Moreover, images lack the structure and grammatical rules of language, and no direct equivalent to NLP tools such as syntactic parsers and regular expression matching exists. In addition, images capture more of the richness of the real world, whereas natural language represents a higher level of abstraction. For example, consider the phrase "red hat" and the multitude of its representations that can be pictured; among these representations, many styles cannot be described in a short sentence. The increasing interest in VQA is driven by the availability of mature techniques in both computer vision and NLP and the availability of relevant large-scale datasets. Therefore, a large body of literature and ground-breaking models on VQA have appeared in the recent five years. The aim of this book is to provide a comprehensive overview of this emerging field, covering fundamental theories, models and datasets, and to suggest promising future directions.

This book can function as a survey of the key models used in the VQA domain and can serve as a textbook for researchers in the computer vision and natural language processing domain, especially researchers and students focusing on visual question answering. We expect our readers to be able to promptly gain knowledge regarding the theory and pros and cons of different popular models in the vision-and-language area. This book can be used as a textbook for a course that introduces the basic principles and models for VQA. Moreover, this book can help students (especially postgraduate students) systematically understand the concepts and methods of vision and language. Through a diverse set of applications and tasks, this book can allow students to explore the use of different models to solve real-world VQA problems. The book is written to be friendly to readers, who require only fundamental machine learning and deep learning knowledge to understand the topics. Basics of deep learning and question answering are first presented to lay the groundwork.

In this book, we first introduce fundamental methods and techniques that are widely used in the vision-and-language domain, including convolutional neural networks, sequential modeling and attention mechanisms. Subsequently, we categorize the VQA tasks into image and video methods according to the visual format. The image-based VQA methods are further classified into five categories, namely, joint embedding, attention mechanism, memory network, compositional models and graph-based models. In addition, we present an overview of other image-based VQA tasks, such as knowledge-based VQA, text-based VQA and visual question generation. In Part III, we discuss video-based visual question answering and its related models. In Part IV, additional VQA-related tasks, including the recently introduced embodied VQA, medical VQA and visual dialogue, which are extensions of VQA, are discussed. Eventually, we highlight future research directions in the VQA field.

Adelaide, Australia   Qi Wu
July 2021

Contents

Chapter 1
Introduction

Abstract Visual question answering (VQA) is a challenging task that has received increasing attention from computer vision, natural language processing and all other AI communities. Given an image and a question in natural language format, reasoning over visual elements of the image and general knowledge are required to infer the correct answer, which may be presented in different formats. In this section, we first explain the motivation behind realizing VQA, i.e., the necessity of this new task and the benefits that the artificial intelligence (AI) field can derive from it. Subsequently, we categorize the VQA problem from different perspectives, including data type and task level. Finally, we present an overview and describe the structure of this book.

1.1 Motivation

The motivation for visual question answering (VQA) [2] arose from image captioning [4, 8, 14, 16, 39, 44], a task originally proposed to connect the computer vision and natural language processing (NLP) fields to examine the image understanding ability and extend the limits of both fields. Figure 1.1 shows examples of image captioning and VQA.

Computer vision (CV) and natural language processing (NLP) are two independent research areas. CV is aimed at teaching machines *how to see* and involves methods for acquiring, processing and understanding images. In contrast, NLP is aimed at teaching machines *how to read* and is focused on enabling interactions between computers and humans in natural language. Both computer vision and NLP belong to the artificial intelligence domain and share similar methods rooted in machine learning.

In recent decades, two fields have witnessed significant advancements toward achieving their respective goals. Moreover, the explosive growth of visual and textual data is driving the combined efforts in these two fields. For example, research on image captioning, i.e., automatic image description [7, 16, 25, 39, 43, 47], has produced powerful methods for jointly learning from image and text inputs to form high-level representations. A popular approach is to combine convolutional

Image Captioning: A group of people enjoying a sunny day at the beach with umbrellas in the sand.

Visual Question Answering:
Q: Why do they have umbrellas? A: Shade.
Q: What is the pattern of the umbrellas? A: Stripe.
Q: Is this a sunny day? A: Yes.
Q: How many stripe umbrellas are here? A: 2
Q: Where is this place? A: Beach
Q: What is in the back? A: Mountains and trees.

Fig. 1.1 Image captioning and visual question answering

Object detection as VQA:
Q: Where is the cat in the image?
A: <x,y,w,h>

Image classification as VQA:
Q: Is there a cat in the image?
A: Yes

Q: Is there a car in the image?
A: No

Fig. 1.2 The task of object detection and image classification can be treated as a VQA problem, except that the question is predetermined

neural networks (CNNs) trained on object recognition with recurrent neural networks (RNNs) to generate word sequences.

In VQA, the machine is presented with an image and a textual question regarding the image. A model must predict the correct answer, typically in the form of a word or a short phrase. Variants include binary (yes/no) [2, 49] and multiple-choice settings [2, 50], in which candidate answers are presented. A closely related task is to "fill in the blank" [48], in which an affirmation describing the image must be completed with one or several missing words. These affirmations essentially amount to questions phrased in declarative form.

A significant distinction between VQA and other tasks in computer vision is that the question to be answered is not determined until run time. In traditional problems such as object detection, classification (Fig. 1.2) and segmentation, the single question to be answered by the algorithm is predetermined, and only the input image changes. A sample object detection question is "Where is the location of XXX in the image?", and a sample classification question is "Are any XXX present?", where "XXX" is an object label. All these questions are predetermined, and the label space is known.

Fig. 1.3 Images returned in a Google search of "a red hat"

In contrast, in VQA, the form of the question is unknown, as is the set of operations required to answer this question. In this sense, this task closely reflects the challenge of general image understanding. VQA is related to the task of natural language (textual) question answering, in which the answer is to be found in a specific textual narrative (i.e., reading comprehension) or large knowledge bases (i.e., information retrieval). In general, textual QA has long been a research focus in the NLP community, and VQA represents its extension to additional visual supporting information. However, VQA is considerably more challenging, as images and videos have a larger number of dimensions and more noise than pure text.

Moreover, images/videos lack the structure and grammatical rules of language, and there is no direct equivalent to NLP tools such as syntactic parsers and regular expression matching. Indeed, as the 2D projection of the physical world, the variants of contents that appear in an image may be unlimited, and it is generally challenging to identify the structure and grammatical rules to define the visual world.

Finally, images capture more of the richness of the real world, and natural language already represents a higher level of abstraction. For example, for the phrase "a red hat", multiple representations can be pictured (Fig. 1.3), several of which cannot be described in a short sentence.

Visual question answering is a significantly more complex problem than image captioning as it frequently requires information not present in the image. This information may range from common sense to encyclopedic knowledge regarding a specific element of the image. In this context, VQA constitutes a truly AI-complete task [2], as it requires multimodal knowledge beyond a single subdomain. Furthermore, after collecting the relevant information, VQA must reason over the information and integrate the supporting facts to derive the answer.

This aspect is a leading reason for the increased interest in VQA, as this task can help evaluate our progress in developing AI systems capable of advanced reasoning combined with deep language and image understanding. Image understanding could, in principle, be evaluated equally well through image captioning. Practically, however, VQA provides a more accessible evaluation metric. In particular, the answers typically contain only a few words. Extended ground-truth image captions are challenging to compare with predicted captions. Although advanced evaluation metrics have been examined, this aspect remains a research problem [11, 22, 38].

Moreover, to realize VQA, superior knowledge representation and reasoning for bridging vision and language are required because to answer questions, both common sense and domain knowledge must be understood. Recently, VQA has been extended to other communities, from medical to robotics. Medical VQA [1, 20] requires the VQA model to answer questions that are related to medical images, such as a CT scan. Moreover, VQA associated with robotics [5] requires an agent to answer questions regarding objects that cannot be seen in the current view, i.e., the agent must navigate to the target location before answering the question, in a framework known as embodied VQA.

One of the first integrations of vision and language is "SHRDLU", which was developed in 1972 [42] and allows users to use language to instruct a computer to move various objects around in a "block world". More recent attempts at creating conversational robotic agents [3, 17, 27, 34] are also grounded in the visual world. However, these frameworks are limited to specific domains and/or restricted language forms. In comparison, VQA addresses free-form open-ended questions. The increasing interest in VQA is driven by the availability of mature techniques in both computer vision and NLP and relevant large-scale datasets. Therefore, a large body of literature on VQA has appeared in recent years, from classical CNN-RNN models to attention mechanisms and transformers.

The trend of VQA models has followed the evolution of deep learning models. The first dominant trend of the VQA model corresponded to the CNN-RNN framework, in which CNN models such as VGG [35] are used to extract the image and video features, and a recurrent neural network (RNN) is subsequently used to harvest features from questions. Next, the visual and textual features are combined and sent to a multi-layer perceptron (MLP) network to predict the answer. Later, different feature combination methods were developed, such as attention-based [46] and bilinear pooling-based [9] methods. With the development of the graph convolutional neural network (GCNN), graphs were introduced in VQA models [29] as they include the structural representation of images. Another line of work focused on the explainability of VQA models, including memory network-based and compositional-based models. In this book, we cover these topics by introducing their theories, advantages and disadvantages.

1.2 Visual Question Answering in AI Tasks

The overall research goal of artificial intelligence is to create technology that allows computers and machines to function in an intelligent manner. The general problem of simulating (or creating) intelligence has been decomposed into several subproblems. These subproblems consist of particular traits or capabilities that researchers expect an intelligent system to display. Several research domains have been explored in the

Fig. 1.4 Artificial
intelligence research goals
and problems. Visual
question answering involves
multiple topics, including
perception, NLP, knowledge
and reasoning (indicated in
red)

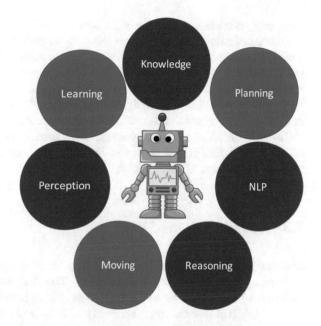

past 50 years, including reasoning, knowledge representation, planning, learning, natural language processing and perception (computer vision), as shown in Fig. 1.4.

Before the VQA, most of the abovementioned topics were separately studied. For example, research on computer vision (CV) is aimed at examining how computers can gain a high-level understanding from digital images or videos. This domain includes methods for acquiring, processing, analyzing and understanding digital images and extracting high-dimensional data from the real world to produce numerical or symbolic information, e.g., in the form of decisions. Classical computer vision tasks include image/video classification [19], object detection and segmentation [23].

In contrast, natural language processing (NLP) is concerned with the interactions between computers and human language; specifically, programming computers require to process and analyze large amounts of natural language data. The relevant tasks cover text and speech processing [30], syntactic analysis, machine translation [45], dialogue management and question answering [32]. Knowledge representation and reasoning (KR&R) is dedicated to representing information regarding the world in a form that a computer system can use to solve complex tasks. The relevant models incorporate findings from logic to automate various kinds of reasoning, such as the application of rules or relations of sets and subsets.

Visual question answering is the first research topic that connects these areas because answering a visual question requires multiple capabilities. First, in contrast to question answering in the NLP domain, which does not include any visual content, questions in VQA are all visually related, i.e., questions pertain to the visual content, such as objects, visual attributes and relationships from images and videos. Hence, VQA requires a machine to understand visual information, which is a typical computer vision task. Second, a VQA model must understand questions that are in the format of natural language. Thus, NLP techniques are required in the VQA task. Third, VQA is a complex task that may require knowledge, for instance, common sense or expertise (such as Wikipedia), to obtain the solution. For example, to clarify whether any mammals are present in an image, the model must know the animals that arc mammals. This knowledge cannot be directly obtained from images or texts and can only be acquired from external knowledge bases.

Overall, the VQA task combines different modalities and sub-AI tasks (such as CV, NLP and KR&R), as shown in Fig. 1.4. Thus, solving the problem of VQA implies the solution of several related AI tasks.

1.3   Categorization of VQA

Modern VQA does not have a long history as other computer vision tasks, such as image classification and segmentation. The first benchmark VQA dataset is DAQUAR [24], which refers to the dataset for question answering on real-world images. DAQUAR, which was proposed in 2014, contains only 795 training and 654 test images. Subsequently, a larger human-annotated dataset named VQA [2] was proposed in 2015. Since then, VQA has emerged as one of the most important topics in computer vision and natural language processing, attracting an increasing number of researchers.

Recently, several datasets and tasks associated with VQA have been proposed, ranging from synthetic images and real images to videos, covering general open questions and medical-related questions. In the following section, we classify the VQA problem in terms of data-source-level or task-level settings (Fig. 1.5).

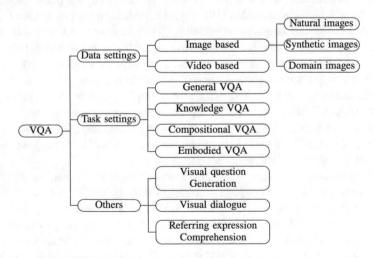

Fig. 1.5 Categorization of VQA according to data and task settings

1.3.1 Classification Based on Data Settings

From the data perspective, we can classify the VQA problem as image-based and video-based VQA.

In image-based VQA, only static images are input to the VQA models, although the images may pertain to different sources. The widely used VQA [2] and VQA 2.0 [10] datasets use images from MS COCO [23] because these images cover rich contextual information. Images from COCOQA [33] and Visual7W [50] have the same source, specifically, MS COCO. The GQA dataset [12] also uses context-rich images selected from the Visual Genome dataset [18].

Other image-based VQA datasets [49] are based on synthetic images. For example, the VQA abstract dataset is based on cartoon images, and the CLEVR dataset [15] is based on synthetic images that cover a range of 3D shapes with different sizes, colors and materials.

Moreover, certain image-based VQA datasets use specific image sources. For example, images from medical VQA tasks [1] are related to the medical field, such as CT, X-ray and ultrasound images. The Text-VQA dataset [36] uses general natural images, but all images include rich text (OCR tokens), and thus, a model must recognize the text that appears in the images. The input of embodied VQA [5] is also images; however, the images are captured from 3D synthetic builds.

Video-based VQA [13, 21, 37] is aimed at answering questions about videos, and it is thus more challenging than image-based VQA. In particular, video-based VQA is considerably different from image-based VQA. First, video-based VQA must manage long sequence images with both rich visual and motion context rather than a single static image. Second, because videos exhibit temporal cues, video-based VQA necessitates additional temporal reasoning abilities to answer questions.

1.3.2 Classification by Task Settings

VQA problems can also be classified according to task settings. The most common setting is to answer general visual questions without providing candidate answers, with the questions pertaining mainly to visual appearance, such as "what color...?", "how many...?" and "what is...?". Many datasets belong to this category, for example, the most widely used VQA [2] and VQA 2.0 [10].

Furthermore, certain VQA questions can only be answered using common sense. KBVQA [40], FVQA [41] and OKVQA [26] belong to this category. In these tasks, although only images and questions are presented during testing, VQA models must query relevant knowledge from the knowledge base to answer questions. For example, for the question "How many mammals are present in the image?", a model must know the animals in the image are mammals. This information can only be obtained from common sense knowledge bases instead of from images.

In addition to knowledge reasoning, several VQA problems are designed to test the compositional reasoning ability of VQA models. For example, the CLEVR dataset [15] designs complex chain and tree-structured reasoning forms and functions and transforms them into natural language questions such as "How many cylinders are in front of the small thing and on the left side of the green object?". Models can answer these questions only if they have strong visual reasoning abilities from both spatial and appearance perspectives.

Embodied VQA [5] represents a slightly different task: an agent is spawned at a random location in a 3D environment and asked a question. The agent must first intelligently navigate to explore the environment, gather information through first-person (egocentric) vision and finally answer the question.

1.3.3 Others

Instead of answering questions, several tasks related to VQA focus on generating questions or maintaining multiple question-answer rounds, known as visual question generation (VQG) [28] and visual dialogue (VisDial) [6], respectively.

VQG [28] can be considered a complementary task of visual question answering. This task involves generating meaningful questions based on the input image. In particular, this task is a multimodal problem involving image understanding and natural language generation, especially using deep learning methods. Certain VQG models use both images and answers to generate questions, while other models use only images.

Visual dialogue [6] is similar to multiple rounds of VQA, requiring an AI agent to hold a meaningful dialog with humans in natural, conversational language regarding the visual content. Given an image, a dialogue history and a question regarding the image, the agent must ground the question in the image, infer context from history and correctly answer the question. Furthermore, another agent must generate a new question based on the dialogue history, thereby maintaining the dialogue. VisDial is adequately disentangled from a specific downstream task to serve as a general test of machine intelligence while being sufficiently grounded in vision to allow the objective evaluation of individual responses and benchmark progress.

Referring expression comprehension [31] is another related topic. In this task, the answer is not text-based but pertains to a detected region. Specifically, this task is aimed at localizing a target object in an image described by a referring expression phrased in natural language. In contrast to the object detection task in which the queried object labels are predefined, the REC problem can only observe the queries during the test. This task has attracted considerable attention from both the computer vision and natural language processing communities, and several lines of work have been proposed, from CNN-RNN models and modular networks to complex graph-based models.

1.4 Book Overview

In this book, we first introduce the fundamental methods and techniques that are widely used in the vision-and-language domain, including convolutional neural networks, sequential modeling and attention mechanisms, in Part I (Chaps. 2 and 3). Subsequently, we divide the VQA tasks into two categories according to the visual format: image and video. The image-based VQA methods (in Part II) are classified into five categories: joint embedding, attention mechanism, memory network, compositional models and graph-based models. Furthermore, we present an overview of other image-based VQA tasks, such as knowledge-based VQA and vision-and-language pretraining for VQA. In Part III, we discuss video-based visual question answering and its related models. In Part IV, we describe additional VQA-related tasks, including the recently highlighted embodied VQA, medical VQA and visual dialogue, which is an extension of VQA. Finally, we discuss future research directions in the VQA field.

This book can function as a survey of the key models used in the VQA domain and be used as a textbook for researchers in the computer vision and natural language processing domain, especially researchers and students focusing on VQA.

References

1. A.B. Abacha, S.A. Hasan, V.V. Datla, J. Liu, D. Demner-Fushman, H. Müller, VQA-Med: overview of the medical visual question answering task at imageclef 2019, in *CLEF (Working Notes)* (2019)
2. S. Antol, A. Agrawal, J. Lu, M. Mitchell, D. Batra, C.L. Zitnick, D. Parikh, VQA: visual question answering, in *Proceedings of the IEEE International Conference on Computer Vision* (2015)
3. R. Cantrell, M. Scheutz, P. Schermerhorn, X. Wu, Robust spoken instruction understanding for HRI, in *2010 5th ACM/IEEE International Conference on Human-Robot Interaction (HRI)* (IEEE, 2010), pp. 275–282
4. X. Chen, C.L. Zitnick, Learning a recurrent visual representation for image caption generation, in *Proceedings of the IEEE Conference on Computer Vision and Pattern Recognition* (2015)
5. A. Das, S. Datta, G. Gkioxari, S. Lee, D. Parikh, D. Batra, Embodied question answering, in *Proceedings of the IEEE Conference on Computer Vision and Pattern Recognition* (2018), pp. 1–10
6. A. Das, S. Kottur, K. Gupta, A. Singh, D. Yadav, J.M. Moura, D. Parikh, D. Batra, Visual dialog, in *Proceedings of the IEEE Conference on Computer Vision and Pattern Recognition* (2017), pp. 326–335
7. J. Donahue, L.A. Hendricks, S. Guadarrama, M. Rohrbach, S. Venugopalan, K. Saenko, T. Darrell, Long-term recurrent convolutional networks for visual recognition and description, in *Proceedings of the IEEE Conference on Computer Vision and Pattern Recognition* (2015)
8. H. Fang, S. Gupta, F. Iandola, R. Srivastava, L. Deng, P. Dollár, J. Gao, X. He, M. Mitchell, J. Platt, et al., From captions to visual concepts and back, in *Proceedings of the IEEE Conference on Computer Vision and Pattern Recognition* (2015)
9. A. Fukui, D.H. Park, D. Yang, A. Rohrbach, T. Darrell, M. Rohrbach, Multimodal compact bilinear pooling for visual question answering and visual grounding. arXiv preprint arXiv:1606.01847 (2016)

10. Y. Goyal, T. Khot, D. Summers-Stay, D. Batra, D. Parikh, Making the V in VQA matter: elevating the role of image understanding in visual question answering, in *Conference on Computer Vision and Pattern Recognition (CVPR)* (2017)
11. M. Hodosh, P. Young, J. Hockenmaier, Framing image description as a ranking task: data, models and evaluation metrics. JAIR 853–899 (2013)
12. D.A. Hudson, C.D. Manning, GQA: a new dataset for real-world visual reasoning and compositional question answering, in *Proceedings of the IEEE/CVF Conference on Computer Vision and Pattern Recognition* (2019), pp. 6700–6709
13. Y. Jang, Y. Song, Y. Yu, Y. Kim, G. Kim, TGIF-QA: Toward spatio-temporal reasoning in visual question answering, in *Proceedings of the IEEE Conference on Computer Vision and Pattern Recognition* (2017), pp. 2758–2766
14. X. Jia, E. Gavves, B. Fernando, T. Tuytelaars, Guiding long-short term memory for image caption generation, in *Proceedings of the IEEE International Conference on Computer Vision* (2015)
15. J. Johnson, B. Hariharan, L. Van Der Maaten, L. Fei-Fei, C. Lawrence Zitnick, R. Girshick, Clevr: a diagnostic dataset for compositional language and elementary visual reasoning, in *Proceedings of the IEEE Conference on Computer Vision and Pattern Recognition* (2017), pp. 2901–2910
16. A. Karpathy, A. Joulin, F.F. Li, Deep fragment embeddings for bidirectional image sentence mapping, in *Proceedings of the Advances in Neural Information Processing Systems* (2014)
17. T. Kollar, J. Krishnamurthy, G.P. Strimel, Toward interactive grounded language acqusition, in *Robotics: Science and Systems* (2013)
18. R. Krishna, Y. Zhu, O. Groth, J. Johnson, K. Hata, J. Kravitz, S. Chen, Y. Kalantidis, L.-J. Li, D. A. Shamma, M. Bernstein, L. Fei-Fei, Visual genome: connecting language and vision using crowdsourced dense image annotations. arXiv preprint arXiv:1602.07332 (2016)
19. A. Krizhevsky, I. Sutskever, G.E. Hinton, Imagenet classification with deep convolutional neural networks, in *Proceedings of the Advances in Neural Information Processing Systems* (2012)
20. J.J. Lau, S. Gayen, A.B. Abacha, D. Demner-Fushman, A dataset of clinically generated visual questions and answers about radiology images. Sci. data **5**(1), 1–10 (2018)
21. J. Lei, L. Yu, M. Bansal, T.L. Berg, TVQA: localized, compositional video question answering. arXiv preprint arXiv:1809.01696 (2018)
22. S. Li, G. Kulkarni, T.L. Berg, A.C. Berg, Y. Choi, Composing simple image descriptions using web-scale n-grams, in *The SIGNLL Conference on Computational Natural Language Learning* (2011)
23. T.-Y. Lin, M. Maire, S. Belongie, J. Hays, P. Perona, D. Ramanan, P. Dollár, C.L. Zitnick, Microsoft COCO: common objects in context, in *Proceedings of the European Conference on Computer Vision* (2014)
24. M. Malinowski, M. Fritz, A multi-world approach to question answering about real-world scenes based on uncertain input, in *Proceedings of the Advances in Neural Information Processing Systems* (2014), pp. 1682–1690
25. J. Mao, W. Xu, Y. Yang, J. Wang, A. Yuille, Deep captioning with multimodal recurrent neural networks (m-RNN), in *Proceedings of the International Conference on Learning Representations* (2015)
26. K. Marino, M. Rastegari, A. Farhadi, R. Mottaghi, OK-VQA: a visual question answering benchmark requiring external knowledge, in *Proceedings of the IEEE/CVF Conference on Computer Vision and Pattern Recognition* (2019), pp. 3195–3204
27. C. Matuszek, N. FitzGerald, L. Zettlemoyer, L. Bo, D. Fox, A joint model of language and perception for grounded attribute learning, in *Proceedings of the International Conference on Machine Learning* (2012)
28. N. Mostafazadeh, I. Misra, J. Devlin, M. Mitchell, X. He, L. Vanderwende, Generating natural questions about an image. arXiv preprint arXiv:1603.06059 (2016)
29. W. Norcliffe-Brown, S. Vafeias, S. Parisot, Learning conditioned graph structures for interpretable visual question answering, in *Proceedings of the Advances in Neural Information Processing Systems* (2018)

30. D. Povey, A. Ghoshal, G. Boulianne, L. Burget, O. Glembek, N. Goel, M. Hannemann, P. Motlicek, Y. Qian, P. Schwarz, et al., The kaldi speech recognition toolkit, in *IEEE 2011 Workshop on Automatic Speech Recognition and Understanding*, number CONF. IEEE Signal Processing Society (2011)
31. Y. Qiao, C. Deng, Q. Wu, Referring expression comprehension: a survey of methods and datasets. IEEE Trans. Multimedia (2020)
32. P. Rajpurkar, R. Jia, P. Liang, Know what you don't know: unanswerable questions for squad. arXiv preprint arXiv:1806.03822 (2018)
33. M. Ren, R. Kiros, R. Zemel, Image question answering: a visual semantic embedding model and a new dataset, in *Proceedings of the Advances in Neural Information Processing Systems* (2015)
34. D. Roy, K.-Y. Hsiao, N. Mavridis, Conversational robots: building blocks for grounding word meaning, in *HLT-NAACL Workshop on Learning Word Meaning from Non-linguistic Data* (Association for Computational Linguistics, 2003), pp. 70–77
35. K. Simonyan, A. Zisserman, Very deep convolutional networks for large-scale image recognition. CoRR arXiv:1409.1556 (2014)
36. A. Singh, V. Natarjan, M. Shah, Y. Jiang, X. Chen, D. Parikh, M. Rohrbach, Towards VQA models that can read, in *Proceedings of the IEEE Conference on Computer Vision and Pattern Recognition* (2019), pp. 8317–8326
37. M. Tapaswi, Y. Zhu, R. Stiefelhagen, A. Torralba, R. Urtasun, S. Fidler, Movieqa: understanding stories in movies through question-answering, in *Proceedings of the IEEE Conference on Computer Vision and Pattern Recognition* (2016), pp. 4631–4640
38. R. Vedantam, C.L. Zitnick, D. Parikh, CIDEr: consensus-based image description evaluation, in *Proceedings of the IEEE Conference on Computer Vision and Pattern Recognition* (2015)
39. O. Vinyals, A. Toshev, S. Bengio, D. Erhan, Show and tell: a neural image caption generator, in *Proceedings of the IEEE Conference on Computer Vision and Pattern Recognition* (2014)
40. P. Wang, Q. Wu, C. Shen, A.v.d. Hengel, A. Dick, Explicit knowledge-based reasoning for visual question answering. arXiv preprint arXiv:1511.02570 (2015)
41. P. Wang, Q. Wu, C. Shen, A.v.d. Hengel, A. Dick, FVQA: fact-based visual question answering. arXiv:1606.05433 (2016)
42. T. Winograd, Understanding natural language. Cogn. psychol. **3**(1), 1–191 (1972)
43. Q. Wu, C. Shen, A.v.d. Hengel, L. Liu, A. Dick, What value do explicit high level concepts have in vision to language problems? in *Proceedings of the IEEE Conference on Computer Vision and Pattern Recognition* (2016)
44. Q. Wu, C. Shen, A.v.d. Hengel, P. Wang, A. Dick, Image captioning and visual question answering based on attributes and their related external knowledge. arXiv preprint arXiv:1603.02814 (2016)
45. Y. Wu, M. Schuster, Z. Chen, Q.V. Le, M. Norouzi, W. Macherey, M. Krikun, Y. Cao, Q. Gao, K. Macherey, et al., Google's neural machine translation system: bridging the gap between human and machine translation. arXiv preprint arXiv:1609.08144 (2016)
46. K. Xu, J. Ba, R. Kiros, A. Courville, R. Salakhutdinov, R. Zemel, Y. Bengio, Show, attend and tell: neural image caption generation with visual attention, in *Proceedings of the International Conference on Machine Learning* (2015)
47. L. Yao, A. Torabi, K. Cho, N. Ballas, C. Pal, H. Larochelle, A. Courville, Describing videos by exploiting temporal structure, in *Proceedings of the IEEE International Conference on Computer Vision* (2015)
48. L. Yu, E. Park, A.C. Berg, T.L. Berg, Visual madlibs: fill in the blank image generation and question answering, in *Proceedings of the IEEE International Conference on Computer Vision* (2015)
49. P. Zhang, Y. Goyal, D. Summers-Stay, D. Batra, D. Parikh, Yin and yang: balancing and answering binary visual questions, in *Proceedings of the IEEE Conference on Computer Vision and Pattern Recognition* (2016)
50. Y. Zhu, O. Groth, M. Bernstein, L. Fei-Fei, Visual7W: grounded question answering in images, in *Proceedings of the IEEE Conference on Computer Vision and Pattern Recognition* (2016)

Part I
Preliminaries

In this part, we introduce the fundamental methods and techniques that are widely used in the vision-and-language domain, including convolutional neural networks, sequential modeling and attention mechanisms.

Chapter 2
Deep Learning Basics

Abstract Deep learning basics are essential for the visual question answering task since multimodal information is usually complex and multidimensional. Therefore, in this chapter, we present basic information regarding deep learning, covering the following: *(1) neural networks, (2) convolutional neural networks, (3) recurrent neural networks and their variants, (4) encoder/decoder structure, (5) attention mechanism, (6) memory networks, (7) transformer networks and BERT, and (8) graph neural networks.*

2.1 Neural Networks

Neural networks are important models in machine learning. The structure of artificial neural networks is similar to that of biological neural networks, which consist of many neurons connected with weighted edges. In this section, we present basic definitions and describe the basic architecture of neural networks.

Neurons are the basic units of neural networks, which take a series of weighted inputs and return a corresponding output. As shown in Fig. 2.1, neuron y calculates the intermediate value x with the weighted sum of the input and bias as $x = \sum_{i=1}^{n} x_i + b$. Next, an ***activation function*** is implemented over x to generate the output of the neuron through $z = f(y)$, which is also the input of the next neuron. The activation function maps a real number to a number between 0 and 1, which represents the activation of the neuron. In the activation function, 0 represents deactivated, while 1 represents fully activated. Several widely used activation functions are the sigmoid function $\sigma(x) = \frac{1}{1+e^{-x}}$, tanh function $tanh(x) = \frac{e^x - e^{-x}}{e^x + e^{-x}}$ and ReLU function $ReLU(x) = 0, if \ x \leq 0 \ ; 1, if \ x > 0$. Additionally, the activation function can be manually designed, following the principles of smoothing and easy calculation.

A neural network usually consists of an input layer, several hidden layers and an output layer, which all contain several neurons. In the input layer, variables $a = \{a_1, a_2, ..., a_n\}$ are input into the neural network; in the hidden layer, calculations are performed; and in the output layer, the output is determined. A simple three-layer fully connected neural network is shown in Fig. 2.2.

© The Author(s), under exclusive license to Springer Nature Singapore Pte Ltd. 2022   15
Q. Wu et al., *Visual Question Answering*, Advances in Computer Vision and Pattern
Recognition, https://doi.org/10.1007/978-981-19-0964-1_2

Fig. 2.1 Neuron

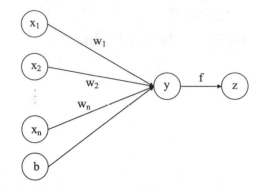

Fig. 2.2 Example of a
simple neural network

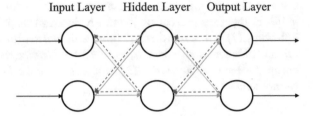

Forward calculation and backpropagation are key steps in the training and testing processes of neural networks . We illustrate the two processes considering the neural network shown in Fig. 2.2.

In the *forward calculation* process, a set of parameters and an input are provided, and the neural network computes the values at each neuron in the forward order, as indicated by the yellow arrows in Fig. 2.2. The generation of outputs of the hidden and output layers, A^2 and A^3, respectively, can be expressed as

$$Z^2 = f(W^1 A^1 + B^2) \quad A^2 = f(Z^2) \quad Z^3 = f(W^2 A^2 + B^3) \quad A^3 = f(Z^3).$$

Here, A^1 is the input vector, W_1 and W_2 represent the learned weighted matrices, B^2 and B^3 are the learned biases and f is the activation function.

Backpropagation, which is an application of the gradient descent of the chain rule, propagates error in the backward order to update the weight matrix, as indicated by the blue dotted arrows in Fig. 2.2. Since the calculation of all the parameters is similar, we consider the backpropagation of weight w_{11}^2 as an example. The corresponding generation can be expressed as follows:

$$\frac{\partial e_{o1}}{\partial w_{11}^2} = \frac{\partial e_{o1}}{\partial a_1^3} \frac{\partial a_1^3}{\partial z_1^3} \frac{\partial z_1^3}{\partial w_{11}^2},$$

$$w_{11}^2 = w_{11}^2 - \eta \frac{\partial e_{o1}}{\partial w_{11}^2}.$$

In addition to the simplest neural network discussed above, a variety of neural network structures exist, which can be classified into several categories: feedforward neural networks, convolutional neural networks, recurrent neural networks and graph neural networks, as described in the following sections.

2.2 Convolutional Neural Networks

A convolutional neural network (CNN) is a kind of multi-layer neural network that can effectively address machine learning problems related to images, especially large images. Through a series of operations, the CNN successfully reduces the high dimensions of the image recognition task with a considerable amount of data. A CNN was first proposed by Yann LeCun [7] and applied to handwritten font recognition (MINST). The structure of the network proposed by LeCun and LeNet is shown in Fig. 2.3.

This network is the most typical convolutional neural network, which consists of convolutional layers, pooling layers and fully connected layers. The convolutional layer cooperates with the pooling layer to form multiple convolutional groups to extract features. Finally, the classification task is accomplished by several fully connected layers. The operations performed by the convolutional layer are inspired by the concept of local receptive fields, and the pooling layer is used to reduce the data dimension.

The local receptive field is designed to lower the number of parameters. It is generally believed that people's cognition of the outside world is from local to global, and the spatial connection of an image is that the local pixels are closely connected to distant pixels. Therefore, each neuron does not need to perceive the global image. Instead, each neuron must perceive only the local part and synthesize the local information at a higher level to obtain the global information, thereby minimizing the number of parameters. The concept of network connection is also inspired by the structure of the biological visual system. In Fig. 2.4, the pictures on the left and right show fully and locally connected neural networks, respectively.

Convolution is similar to a sliding window, and the convolution kernel and corresponding image pixels are multiplied and summed. The parameters of the convolution kernel remain the same when convolution is implemented over a data matrix, which enables weight sharing and decreases the number of parameters. To generate more features over the same data, multiple convolution kernels can be applied to repeat the convolution over a data matrix.

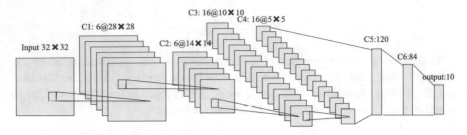

Fig. 2.3 Network structure of LeNet

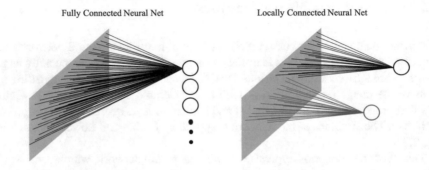

Fig. 2.4 Different connections

2.3 Recurrent Neural Networks and Variants

A recurrent neural network (RNN) is used to process sequential data with inner relations. By processing the data sequentially, an RNN can memorize the data that it has processed. The structure of an RNN is shown in Fig. 2.5. At timestamp t, the RNN operates on data x_t with the hidden state s_{t-1} at timestamp $t - 1$. The hidden state h_t at timestamp t and output y_t at timestamp t are generated by $y_t, h_t = f(x_t, h_{t-1})$. Therefore, the final output $O = \{y_1, y_2, y_3 ... y_T\}$ is a sequence containing all the outputs at each timestamp.

Long short-term memory (LSTM) is a special variant of RNN, aiming to solve the problem of gradient disappearance and gradient explosion during long sequence training. The LSTM can outperform ordinary RNNs in the case of long sequences. As depicted in Fig. 2.6, an LSTM has two states, h_t and c_t, in contrast to RNNs, which have only one state. First, z, z^i, z^f, z^o are generated by Eq. (2.1) as follows:

$$
\begin{aligned}
z &= tanh(W[x^t, h^{t-1}]), \\
z^i &= \sigma(W^i[x^t, h^{t-1}]), \\
z^f &= \sigma(W^f[x^t, h^{t-1}]), \\
z^o &= \sigma(W^o[x^t, h^{t-1}]),
\end{aligned}
\tag{2.1}
$$

where W is the weighted matrix, and operator $[a, b]$ means concatenation of two matrices on the y-axis. Subsequently, cell state c_t, hidden state h_t and output y_t are generated by z, z^i, z^f, z^o from Equation (2.2).

$$c^t = z^f \odot c^{t-1} + z^i \odot z,$$
$$h^t = z^o \odot tanh(c^t), \tag{2.2}$$
$$y^t = \sigma(W'h^t).$$

In this manner, the LSTM controls the hidden states by the cell state, thereby remembering the long-term memory and forgetting the unimportant information. This framework is efficient for tasks requiring long-term memory. However, the presence of a larger number of parameters renders training more difficult. Therefore, we usually tend to use GRUs that have the same effect as LSTM but with fewer parameters.

The gate recurrent unit (GRU) is another special variant of RNN, with the same aim as that of the LSTM network but fewer parameters; consequently, the GRU is considerably easier to train. The structure of GRUs is shown in Fig. 2.7.

Similar to a basic RNN, a GRU takes the hidden state of the last timestamp h^{t-1} and x^t as input, generating two gates r and z according to Eq. (2.3). Subsequently, hidden state h_t and output y_t are generated using Eq. (2.4).

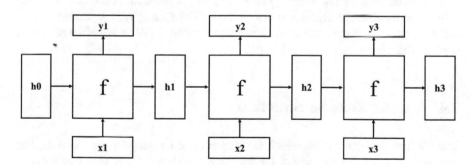

Fig. 2.5 RNN

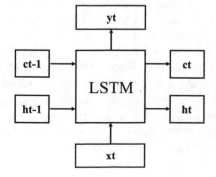

Fig. 2.6 LSTM

Fig. 2.7 GRU

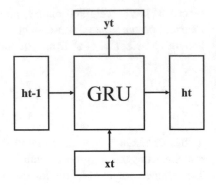

$$r = \sigma(W^r[x^r, h^{t-1}]),$$
$$z = \sigma(W^z[x^t, h^{t-1}]).$$
(2.3)

$$h^{'} = tanh(W[x^t, h^{t-1} \odot r]),$$
$$h^t = (1 - z) \odot h^{t-1} + z \odot h^{'},$$
$$y^t = \sigma(W^o h^t).$$
(2.4)

Compared with LSTM, there is one less "gate" in the GRU; thus, the GRU can achieve performance equivalent to that of the LSTM with fewer parameters. Given the computational ability of the hardware and time cost, GRUs are often preferred over LSTM.

2.4 Encoder/Decoder Structure

The encoder/decoder model, which is composed of an encoder and decoder, has emerged as a popular alternative for machine translation tasks [3]. The encoder encodes a sequence of inputs into an intermediate code, and the decoder decodes the intermediate code and generates the output sequence. The basic pipeline of the encoder/decoder model is shown in Fig. 2.8 and can be described as follows. The input sequence is described as $X = x_1, x_2, \cdots, x_T$. At each timestamp t, hidden state h_t is generated as $h_t = RNNs(x_t, h_{t-1})$, where h_{t-1} is the hidden state at timestamp $t - 1$. After all the hidden states have been generated, the result $c = f(h_1, h_2, ..., h_T)$, where c is a contextual representation related to the whole sequence. In the decoding process, the decoder outputs the prediction y_t at timestamp t according to the probability function $p(y_t|y_1, t_2, ..y_{t-1}, c) = g(y_{t-1}, s_t, c)$, where y_{t-1} is the output at timestamp $t - 1$, and the hidden state of the decoder RNN unit, s_t, is generated as $s_t = RNNs(s_{t-1}, y_{t-1})$.

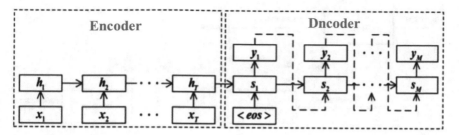

Fig. 2.8 Encoder/decoder structure

2.5   Attention Mechanism

The attention mechanism was first used in the field of natural language processing for machine translation [2] and later applied in the field of image processing. The attention mechanism can be described as a function that maps a query and a set of key-value pairs to the output [13]. Specifically, the framework can be defined as $o = Attention(Q, K, V)$, where Q is the query, K is the key, V represents the values and O is the output. The query and source, which consist of key-value pairs, are inserted in the attention model to generate the output. First, a score function is used to compute the similarity between the query and key as $s_i = Score(query, key_i)$, with the query used as a reference or guide to find the related key-value pairs. Subsequently, an alignment function is operated over scores of all the keys by $a_i = \frac{exp(s_i)}{\sum exp(s_i)}$. Finally, the weighted value, also known as query-guided source representation, is calculated as the result of attention by $c = \sum_i a_i value_i$.

2.6   Memory Networks

Memory networks [15] (MemNNs) were proposed by Facebook AI in 2015 to use memory components to store information to realize the memorization of long-term memory. Many existing neural network models, including RNN, LSTM and GRU, can memorize sequential information to a certain degree; however, these memories are often inadequate. A MemNN model consists of memory m and four components:

- I, input feature map, which converts the incoming input to the internal feature representation.
- G, generalization module, which updates old memories given the new input.
- O, output feature map, which produces a new output.
- R, response, which converts the output into the desired response format.

In the VQA task, the memory array can store the entire visual sequential informa-tion and can thus retain the long-term memory of the sequence as a knowledge base. In each updated timestamp t, the memory array indexed by m_i in the MemNN can

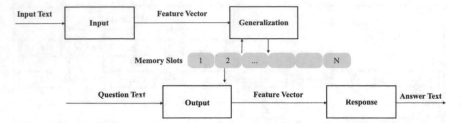

Fig. 2.9 Structure of a memory network

be updated as $m_i = G(m_i, I(x), m), \in i$. Subsequently, the memory can be used to infer the answer to a given question. According to the query and current memory array, the output feature map unit conducts an inference to obtain the contextual representation in the feature space. Finally, the response unit converts the contextual representation into the predicted answer (Fig. 2.9).

MemNN frameworks involve a model in combination with several components; therefore, they cannot be trained in an end-to-end manner. Sukhbaatar [12] proposed end-to-end memory networks (MemN2Ns) based on the MemNN, which can be trained end-to-end, and established a process for the repetitive extraction of useful information to achieve multiple inferences. One characteristic of MemN2N is that it uses two memory arrays to store the converted input sequences, denoted as input memory m^v and output memory m^o. The input memory m^v is generated using an embedding matrix A to convert the input sequence, and the output memory m^o is generated using another embedding matrix B to convert the input sequence. Subsequently, the weight of the embedded question over input memory m^v is calculated and used to determine the weighted output memory c. Finally, the output c and embedded question u are introduced in the softmax function to generate the answer. The MemN2N model can be trained iteratively by updating the question embedding per iteration.

Dynamic memory networks (DMNs) [6] are improved versions of MemNN. The network consists of the following four components:

- Input module
- Question module
- Episodic memory module
- Answer module

The input module and question module use GRU to generate encoded visual representations $c = \{c1, c2, ..., c_T\}$ and textual representations q. The episodic memory module generates contextual representations with an attention mechanism and a memory update mechanism. For the ith iteration, the gate value is produced as $g_t^i = G(c_t, m_{i-1}, q)$, where G is the gate function and m_{i-1} is the memory generated in the last iteration. The gate value is used to generate episode e^i as follows:

$$h_t^i = g_t^i GRU(c_t, h_{t-1}^i) + (1 - g_t^i)h_{t-1}^i,$$

Fig. 2.10 Structure of dynamic memory networks

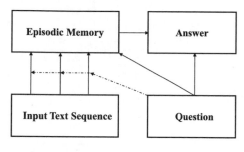

$$e^i = h^i_{last}.$$

The episodic memory module uses a GRU to update the episodic memory: $m^i = GRU(e^i, m^{i-1})$. Finally, the answer module uses the episodic memory to predict the answer vector, which is used as the initial state of the GRU in the next iteration (Fig. 2.10).

In conclusion, the memory array of the DMN is dynamic during the process of reasoning, while the memory arrays of MemNN and MemN2N are static [13].

2.7 Transformer Networks and BERT

A transformer framework, proposed by Google [14] in 2017, is a type of seq2seq model that replaces LSTM with an attention structure. The structure of the transformer, as shown in Fig. 2.11, consists of an encoder and a decoder.

The encoder contains six layers depicted as Nx on the left figure, which consist of two sublayers, including a multihead self-attention mechanism and a fully connected feedforward network. The multihead self-attention mechanism can be expressed as follows:

$$MultiHead(Q, K, V) = Concat(head_1, head_2, ...head_h)W^o$$

$$head_i = Attention(QW^Q_i, KW^K_i, VW^V).$$

Here, Q, K, V are the same as those for the self-attention mechanism. A positionwise feedforward network operates a nonlinear function over the input.

The decoder is similar to the encoder, except an additional sublayer of attention is added to the decoder. The decoder takes the output of the encoder and output of the last position as the input. The second multihead attention in the decoder is different from that of the encoder (a multihead self-attention), in which the key and query are the output of the encoder, and the query is the output of the last position.

Before the data are input to the encoder and decoder, they are subjected to positional encoding, and sequential data embedding is introduced with positional embedding, which can be defined as follows:

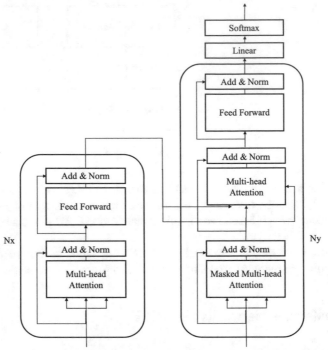

Positional Encoding Added Embedding Input   Positional Encoding Added Embedding Output

Fig. 2.11 Structure of transformer networks

$$PE_{pos,2i} = sin(pos/10000^{2i/d_{model}})$$

$$PE_{pos,2i+1} = cos(pos/10000^{2}i/d_{model}).$$

Bi-directional encoder representation from transformers (BERT) [5] is a pre-trained model proposed by Google AI in 2018, which is built upon a bi-directional transformer encoder block. The training of the BERT can be characterized by pre-training, a deep structure, a bidirectional transformer and language understanding.

2.8   Graph Neural Networks

Graph neural networks (GNNs) were proposed by Scarselli et al. [11] to extend the existing neural networks for processing graph-structured data. In particular, GNNs are aimed at learning a state embedding, which encodes the information of the neighborhood for all the nodes. Later, state embedding is used to produce an output, such as the distribution of the predicted node label.

First, we introduce several basic definitions. A graph, denoted as G, is represented as $G = (V, E)$, where $V = \{v_i\}$ is the set of nodes and $E = \{e_{ij}|e_{ij} = (v_i, v_j)\}$ is the set of edges. The adjacency matrix A is an $n \times n$ matrix with $A_{ij} = 1$ if $e_{ij} \in E$ and $A_{ij} = 0$ if $e_{ij} \notin E$. The nodes can have attributes $X = \{x_i\}$, and the edges can have attributes $X^e = \{x_i^e\}$. A spatial-temporal graph, denoted as $G^{(t)}$, is an attributed graph in which the node attributes change dynamically over time and can be represented as $G^{(t)} = (V, E, X^t)$, where X^t represents time-related information.

In this chapter, we introduce the definition and frameworks of three types of GNNs, including vanilla graph neural networks (GNNs), recurrent graph neural networks (RecGNNs) and convolutional graph neural networks (ConvGNNs).

Vanilla graph neural networks (GNNs) were proposed by Scarselli et al. [11] to address the indicted attributed homogeneous graph. The model learns the node embedding h_v as $h_v = f(x_v, x_{co[v]}, h_{ne[v]}, x_{ne[v]})$, where f is a parametric function, x_v is the attribution of node v, $x_{ne[v]}$ is the attribution of the neighboring nodes of node v, $h_{ne[v]}$ is the embedding of the neighboring nodes of node v and $x_{co[v]}$ is the attribution of the connected edges of node v. The output embedding of node v is generated as $o_v = g(h_v, x_v)$, where g is a parametric function. Let H, O, X represent the matrices constructed by stacking all the states, outputs and features; we can iteratively update the node embedding via $H^{t+1} = F(H^t, X)$. The loss of the vanilla GNN can be expressed as $loss = \sum_{i=1}^{p}(t_i - o_i)$, where t_i is the target output of node i. Although the vanilla GNN can effectively manage graph-structured data, certain limitations remain, including inefficient updates, use of the same parameters in the iterations and ineffective use of the features of the edges.

Recurrent graph neural networks (RecGNNs) use gate mechanisms from RNNs such as GRUs and LSTMs in the propagation step to alleviate the restrictions of the vanilla GNN model and enhance the effectiveness of long-term information propagation across the graph. Since RecGNNs have many variants, we introduce only the basic framework of the RecGNNs referring to GGNN [8]. In iteration t, we update the embedding of node v as follows:

$$a_v^t = A_v^T [h_1^{t-1}, ..., h_N^{t-1}]^T + b,$$

$$z_v^T = \sigma(W^z a_v^t + U^z h_v^{t-1}),$$

$$r_v^t = \sigma(W^r a_v^t + U^r h_v^{t-1}),$$

$$\widetilde{h_v^t} = tanh(W a_v^t + U(r_v^t \odot h_v^{t-1})),$$

$$h_v^t = (a - z_v^t) \odot h_v^{t-1} + z_v^t \odot \widetilde{h_v^t}.$$

A_v is the submatrix of the graph adjacency matrix A and denotes the connection of node v with its neighbors. The GRU-like update functions take information from the neighbors of each node and form the previous iteration to generate a new embedding.

Graph convolutional networks (GCNs) generalize convolutions to the graph domain, thereby defining the convolution operation on graphs. GCNs and their variants can be categorized as spectral approaches, which work with a spectral representation of the graphs, and spatial approaches, which define convolutions directly on the graph, thereby operating on spatially close neighbors [9]. Herein, we introduce only the classical framework of the GCNs of spatial approaches. First, we define the receptive field of nodes. Second, graph normalization is performed on the graph to specify the order of the nodes in the receptive field. Finally, a CNN architecture can be used, considering the normalized neighborhoods as receptive fields and the node and edge attributes as channels [10].

References

1. V. Badrinarayanan, A. Kendall, R. Cipolla, Segnet: a deep convolutional encoder-decoder architecture for image segmentation. IEEE Trans. Pattern Anal. Mach. Intell. **39**(12), 2481–2495 (2017)
2. D. Bahdanau, K. Cho, Y. Bengio, Neural machine translation by jointly learning to align and translate. arXiv preprint arXiv:1409.0473 (2014)
3. K. Cho, B. Van Merriënboer, D. Bahdanau, Y. Bengio, On the properties of neural machine translation: encoder-decoder approaches. arXiv preprint arXiv:1409.1259 (2014)
4. K. Cho, B. Van Merriënboer, C. Gulcehre, D. Bahdanau, F. Bougares, H. Schwenk, Y. Bengio, Learning phrase representations using rnn encoder-decoder for statistical machine translation. arXiv preprint arXiv:1406.1078 (2014)
5. J. Devlin, M.-W. Chang, K. Lee, K. Toutanova, Bert: pre-training of deep bidirectional transformers for language understanding. arXiv preprint arXiv:1810.04805 (2018)
6. A. Kumar, O. Irsoy, P. Ondruska, M. Iyyer, J. Bradbury, I. Gulrajani, V. Zhong, R. Paulus, R. Socher, Ask me anything: dynamic memory networks for natural language processing, in *International Conference on Machine Learning* (PMLR, 2016), pp. 1378–1387
7. Y. LeCun, L. Bottou, Y. Bengio, P. Haffner, Gradient-based learning applied to document recognition. Proc. IEEE **86**(11), 2278–2324 (1998)
8. Y. Li, D. Tarlow, M. Brockschmidt, R. Zemel, Gated graph sequence neural networks. arXiv preprint arXiv:1511.05493 (2015)
9. Z. Liu, J. Zhou, Introduction to graph neural networks. Synth. Lect. Artif. Intell. Mach. Learn. **14**(2), 1–127 (2020)
10. M. Niepert, M. Ahmed, K. Kutzkov, Learning convolutional neural networks for graphs, in *International Conference on Machine Learning* (PMLR, 2016), pp. 2014–2023
11. F. Scarselli, M. Gori, A.C. Tsoi, M. Hagenbuchner, G. Monfardini, The graph neural network model. IEEE Trans. Neural Netw. **20**(1), 61–80 (2008)
12. S. Sukhbaatar, A. Szlam, J. Weston, R. Fergus, End-to-end memory networks. arXiv preprint arXiv:1503.08895 (2015)
13. G. Sun, L. Liang, T. Li, B. Yu, B. Wu, B. Zhang, Video question answering: a survey of models and datasets. Mobile Netw. Appl. 1–34 (2021)
14. A. Vaswani, N. Shazeer, N. Parmar, J. Uszkoreit, L. Jones, A.N. Gomez, Ł. Kaiser, I. Polosukhin, Attention is all you need, in *Advances in Neural Information Processing Systems* (2017), pp. 5998–6008
15. J. Weston, S. Chopra, A. Bordes, Memory networks. arXiv preprint arXiv:1410.3916 (2014)

Chapter 3
Question Answering (QA) Basics

Abstract The main objective of the question answering (QA) task is to provide relevant answers in response to questions asked in natural language through either a prestructured database or a collection of natural language documents [11]. The basic architecture usually consists of three components: a question processing unit, a document processing unit and an answer processing unit. The question processing unit first analyzes the structure of the given question and transforms the question into a meaningful format compatible with the QA domain. The document processing unit generates a dataset or a model that provides information for answer generation. The answer processing unit extracts the answer from information and formatted questions. In this chapter, we discuss the QA task from the following aspects: rule-based methods, information retrieval-based methods, neural semantic parsing-based methods and approaches taking knowledge base into account.

3.1 Rule-Based Methods

Rule-based methods usually implement handmade rules to identify either the expected answer types or documents. These handmade rules may be accurate but time-consuming to acquire and are usually used in language processing.

Riloff et al. [10] developed a rule-based system Quarc to answer a given question with a short story. Quarc identifies the type of question (e.g., who/what/when/where/why) and uses a separate set of rules for each question type. Partial parser Sundance is applied to each sentence in the story and the question to obtain the morphological analysis, part-of-speech tagging, semantic class tagging and entity recognition from it to generate scores for sentences on which rules are applied. Rules award a certain number of points to a sentence, and herein, we consider rules for the *who* questions as an example, as shown in Fig. 3.1. The confident and good clue are different constants for points, Q is the question, S is the sentence in the story and NAME is a person-noun that contains at least one human word. The answer is the sentence with the highest score.

Based on this system, Gusmita et al. [5] proposed a rule-based system that uses both relevant documents and rules. First, the relevant documents toward the keywords

Fig. 3.1 Rules for
WHO-type question

1. Score(S) += WordMatch(Q,S)
2. If not contains(Q,NAME) and contains(S,NAME)
Then Score(S) += confident
3. If not contains(Q,NAME) and contains(S,name)
Then Score(S) += good_clue
4. If Contains(S,{NAME,HUMAN})
Then Score(S) += good_clue

are gathered and subsequentially used to identify answer candidates by using a rule-based method. The system builds its own rules for Indonesian translation.

Archana et al. [1] proposed a rule-based system to identify the answer having the same Vibhakthi and POS attributes as the question. The system first analyzes the question given by the user and identifies the question type (e.g., who/which/whom/how much, etc.). Subsequently, the rule-based analysis of Malayalam, including Pos, Vibhakthi and Sandhi analyses, is performed upon questions to extract question features. The analysis document corpus with the same rule is used to find the answer that best matches the question.

3.2   Information Retrieval-Based Methods

Rule-based methods, described in the last section, have many limitations since the manual rules are usually difficult to prepare for complex problems. Information retrieval-based methods in which the relevant contexts for each question/answer candidate pair are extracted using an information retrieval approach are more widely used in question answering tasks.

Sebastian et al. [9] proposed a two-step method that combines information retrieval techniques optimized for question answering with deep learning models for natural language inference on multiple-choice questions. In the first step, relevant knowledge support is extracted using Lucene, taking question/answer candidate pairs. Subsequently, improved semantic similarity computation is performed over the tuple (question, answer and context), which predicts whether the current answer is the correct one. The solver employs a bi-directional attention flow (BiDAF) architecture to generate the answer. The rule-based methods introduced in the last section have many limitations since the manual rules are usually difficult to prepare for complex problems. Information retrieval-based methods in which the relevant contexts for each question/answer candidate pair are extracted using an information retrieval approach are more widely used in question answering tasks.

Manna et al. [8] proposed an information retrieval-based QA system to match recipe-related information with food. The QA system consists of the following four modules:

- Apache Lucene module, which serves as a retrieval database related to cooking recipe information and cooking-related documents.
- Query processing module, which tokenizes the question, identifies the question type and extracts the segments of information from the question, including POS tags.
- Document processing module, which obtains relevant information from one or more data systems, and sorts and organizes the obtained documents into the Apache Lucene module.
- Answer processing module, which checks the information file and provides precise answers for a specific question.

3.3 Neural Semantic Parsing for QA

Semantic parsing is the problem of translating human language into computer language and is therefore at the heart of the question answering task, in which the question and answer are represented in a human language format. The encoder/decoder architecture neural network, described previously, is usually used for semantic parsing.

The semantic parsing for questions usually involves tokenization and relation extraction to identify the relation and entities. Yih et al. [12] performed semantic parsing on a single-relation question. First, the question was separated into two disjoint parts: the entity mention and relation pattern. Later, a convolutional neural network-based semantic model (CNNSM), consisting of a hashing layer, convolutional layer and max pooling layer, took word sequences from the entity mentions and relation patterns to generate semantic embeddings. Two CNN semantic models were trained for the pattern relation and mention entity pairs separately, and the semantic relevance score was defined as the cosine score of the two semantic vectors. In this manner, the semantic relevance score of the pattern and entity in the knowledge with relation and mention in the question was calculated to generate the answer.

Krishnamurthy et al. [7] proposed a semantic parsing model that followed an encoder/decoder architecture using recurrent neural networks with long short-term memory (LSTM) cells. Two modifications were introduced in the encoder/decoder architecture. First, the encoder included a special entity embedding and linking module that produced a link embedding for each question token that represented the table entities to which it links. Second, type-constrained grammar was added to ensure that the generated logical forms satisfied the type constraints.

3.4 Knowledge Base for QA

A knowledge base (KB) is a technology used to store complex structures and unstructured information used by a computer system, each piece of knowledge of which is represented by a triple containing two entities and the relation between entities. Two types of KBs exist: curated KBs and extracted KBs. Curated KBs extract a large number of entities and entity relationships from web-based knowledge bases such as Wikipedia and WordNet, which can be understood as structured Wikipedia, including Freebase [2] and Yago2 [6]. Extracted KBs extract knowledge directly from the internet, including Open Information Extraction (OpenIE) [4] and Never-Ending Language Learning (NELL) [3]. Compared with curated KBs, the knowledge in extracted KBs is usually more diverse and in a natural language manner, with more noise and lower accuracy.

The two key technologies in generating a knowledge base are entity linking and relation extraction. Entity linking connects the entity name in the document to a specific entity in the knowledge base. This process encounters two problems in the field of natural language processing: entity recognition and entity disambiguation. Relation extraction extracts the entity relationship in the document through key technologies such as part-of-speech tagging (POS), syntax analysis, dependency tree generation, construction of SVM and maximum entropy model and classification of relationships.

References

1. S.M. Archana, N. Vahab, R. Thankappan, C. Raseek, A rule based question answering system in malayalam corpus using vibhakthi and pos tag analysis. Procedia Technol. **24**, 1534–1541 (2016)
2. K. Bollacker, C. Evans, P. Paritosh, T. Sturge, J. Taylor, Freebase: a collaboratively created graph database for structuring human knowledge, in *Proceedings of the 2008 ACM SIGMOD International Conference on Management of Data* (2008), pp. 1247–1250
3. A. Carlson, J. Betteridge, B. Kisiel, B. Settles, E.R. Hruschka, T.M. Mitchell, Toward an architecture for never-ending language learning, in *Twenty-Fourth AAAI Conference on Artificial Intelligence* (2010)
4. O. Etzioni, M. Banko, S. Soderland, D.S. Weld, Open information extraction from the web. Commun. ACM **51**(12), 68–74 (2008)
5. R.H. Gusmita, Y. Durachman, S. Harun, A.F. Firmansyah, H.T. Sukmana, A. Suhaimi, A rule-based question answering system on relevant documents of indonesian quran translation, in *2014 International Conference on Cyber and IT Service Management (CITSM)* (IEEE, 2014), pp. 104–107
6. J. Hoffart, F. Suchanek, K. Berberich, E. Kelham, G. de Melo, G. Weikum, F. Suchanek, G. Kasneci, M. Ramanath, A. Pease, Yago2: a spatially and temporally enhanced knowledge base from wikipedia. Commun. ACM **52**(4), 56–64 (2009)
7. J. Krishnamurthy, P. Dasigi, M. Gardner, Neural semantic parsing with type constraints for semi-structured tables, in *Proceedings of the 2017 Conference on Empirical Methods in Natural Language Processing* (2017), pp. 1516–1526
8. R. Manna, D. Das, A. Gelbukh, Information retrieval-based question answering system on foods and recipes, in *Mexican International Conference on Artificial Intelligence* (Springer, 2020), pp. 260–270
9. G.-S. Pîrtoacă, T. Rebedea, Ş. Ruşeţi, Improving retrieval-based question answering with deep inference models, in *2019 International Joint Conference on Neural Networks (IJCNN)* (IEEE, 2019), pp. 1–8

10. E. Riloff, M. Thelen, A rule-based question answering system for reading comprehension tests, in *ANLP-NAACL 2000 Workshop: Reading Comprehension Tests as Evaluation for Computer-Based Language Understanding Systems* (2000)
11. M.A.C. Soares, F.S. Parreiras, A literature review on question answering techniques, paradigms and systems. J. King Saud Univ.-Comput. Inf. Sci. **32**(6), 635–646 (2020)
12. W. Yih, X. He, C. Meek, Semantic parsing for single-relation question answering, in *Proceedings of the 52nd Annual Meeting of the Association for Computational Linguistics (Volume 2: Short Papers)* (2014), pp. 643–648

Part II
Image-Based VQA

Visual question answering methods can be distinguished based on different input visual formats. In this part, we focus on the classic VQA problem, specifically, image-based VQA, which accepts only images as the input. This part includes three chapters that describe classical VQA methods, knowledge-based VQA and vision-and-language pretraining methods.

Chapter 4
Classical Visual Question Answering

Abstract VQA has received considerable attention from both the computer vision and the natural language processing research communities in recent years. Given an image and the corresponding question in natural language, a VQA system is required to comprehend the question and find the essential visual elements in the image to predict the correct answer. In this chapter, we first introduce the prevalent datasets for VQA tasks, such as the COCO-QA, VQA v1 and VQA v2 datasets. Subsequently, we present a detailed description of several classical VQA methods classified as joint embedding methods, attention-based methods, memory networks and compositional methods.

4.1 Introduction

Many variants and extended versions of visual question answering (VQA) have been proposed since its popularization. For example, video question answering [36, 46] extends the VQA from images to videos, TextVQA [33] requires VQA models to answer optical character recognition (OCR)-related questions and knowledge-based VQA [37, 38] aims to answer knowledge-related visual questions. These extensive and advanced topics are introduced and discussed in Parts III and IV and the following chapter. In this chapter, we focus on the classical VQA task with the corresponding datasets and methods.

In the most common form of visual question answering (VQA), a computer is presented with an image and a textual question regarding the image (see the example in Fig. 4.1). Subsequently, the computer must determine the correct answer, typically in the form of a few words or a short phrase. Several other variants of the answer space exist, for example, binary answers (yes/no) [4, 48] and multiple-choice settings [4, 51], in which candidate answers are presented.

In this chapter, we present a comprehensive review of classical VQA methods classified into five categories based on the nature of their main contributions. Incremental contributions mean that most methods belong to multiple categories.

First, *joint embedding approaches* (Sect. 4.4) are motivated by the advancements in deep neural networks in both the computer vision and NLP domains. These methods use convolutional and recurrent neural networks (CNNs and RNNs, respectively) to learn the embeddings of images and sentences in a common feature space. These entities can be fed to a classifier that predicts an answer [11, 25, 27].

Second, *attention mechanisms* (Sect. 4.5) improve on the abovementioned method by focusing on specific parts of the input (image and/or question). Attention in VQA [1, 3, 7, 14, 40, 43, 51] has been inspired by the success of similar techniques in the context of image captioning [41], the main idea of which is to replace holistic (image-wide) features with spatial feature maps and allow for interactions between the question and specific regions of these maps. The famous transformer [34] model is an extensive version of the attention mechanism. In this section, the attention mechanism is explained, and different variants are discussed.

Third, *memory networks* (Sect. 4.6) extend the attention mechanism but allow a model to read and write operations on an internal representation of the input—in our case, the question and image.

Fourth, *compositional models* (Sect. 4.7) allow us to tailor the performed computations to each problem instance. For example, Andreas et al. [3] used a parser to decompose a given question and built a neural network from modules with a composition that reflects the structure of the question.

Finally, *graph neural networks* (Sect. 4.8) enable models to perform reasoning on structure representations, such as a scene graph representation. This type of model exhibits excellent performance, especially in terms of spatial and logical reasoning. Graph attention is later introduced to boost performance.

Furthermore, in this chapter, we examine datasets available for training and evaluating VQA systems. These datasets vary widely along three dimensions, specifically, in terms of (i) their size, i.e., the number of images, questions and different concepts represented; (ii) the amount of required reasoning, e.g., whether the detection of a single object is sufficient or whether inference is required over multiple facts or concepts; and (iii) synthetic or human annotation.

4.2   Datasets

A number of datasets have been proposed for the research on VQA. These datasets contain, at a minimum, triples made of an image, a question and its correct answer. Additional annotations are provided in certain cases, such as image captions, image regions supporting the answers or multiple-choice candidate answers. Datasets and questions within the datasets vary widely in their complexity, the amount of reasoning and nonvisual (e.g., "common sense") information required to infer the correct answer. This section presents a comprehensive comparison of the available datasets and discusses their suitability for evaluating different aspects of VQA systems. We only focus on the general classical VQA dataset in this section. Other VQA datasets for specific domains, such as Medical VQA, TextVQA and knowledge-based VQA, are introduced in other parts.

A key characteristic that differentiates the various datasets is the type of their images, which we broadly classify into natural, clip art and synthetic. Datasets widely used in the initial stage, such as DAQUAR [26], COCO-QA [32] and VQA-v1-real [4], use natural (i.e., real) images. The most widely used dataset at present, specifically, the VQA-v2 [13] dataset, an extensive version of the original VQA-v1-real, also uses natural images. The VQA-v1-abstract [4] and its balanced version [48] are based on synthetic clip art (i.e., cartoon) images.

The second key difference between datasets is the question/answer format: open-ended *versus* multiple-choice questions. The former case does not include any predefined set of answers and commonly pertains to DAQUAR, COCO-QA, FM-IQA [11] and Visual Genome [19]. The multiple-choice setting provides a limited set of possible answers to each question and is used, for example, in Visual Madlibs [45]. The VQA-v1-real and Visual7W [51] datasets allow evaluation with either open-ended or multiple-choice questions. The results from the two settings cannot be compared, and the open-ended setting is considered more challenging to quantitatively evaluate. Most authors address the VQA-v1-real dataset in the open-ended setting, while the authors of Visual7W recommend the multiple-choice setting for a more interpretable evaluation.

Details of these datasets are presented in the following text, and the key characteristics are summarized in Table 4.1.

DAQUAR

DAQUAR [26], which stands for dataset for question answering on real-world images, is the first dataset proposed for VQA tasks. DAQUAR is constructed based on the NYU-Depth v2 dataset and contains 1,449 images (795 images for training and 654 images for testing). The corresponding question/answer pairs are collected in two ways: *synthetic*, in which question/answer pairs are automatically generated according to annotations in the NYU dataset by a predefined template, and *human*, in which question/answer pairs are collected by human annotators to focus on basic colors, numbers, objects and sets. In total, 12468 QA pairs are collected, of which 6,794 are used for training and 5,674 are used for testing. DAQUAR is the first large VQA dataset, which promoted the development of early VQA methods. However, its disadvantage lies in the restriction of answers and strong biases toward a few objects.

COCO-QA

COCO-QA [32] is constructed based on the Microsoft Common Objects in Context data (COCO) dataset [22] and contains 123,287 images (72,783 for training and 38,948 for testing). The corresponding question/answer pairs are collected in an automatic manner, with the QA pairs generated by turning the image descriptions into the question/answer form. Each image in the COCO-QA has one question/answer pair. COCO-QA increases the training data for VQA tasks; however, the automatically generated questions have a high repetition rate.

Table 4.1 Major datasets for VQA and their main characteristics

Dataset	Source of images	Number of images	Number of questions	Num. questions/Num. images	Num. question categories	Question collection	Average quest. length	Average ans. length	Evaluation metrics
DAQUAR	NYU-Depth V2	1,449	12,468	8.6	4	Human	11.5	1.2	Acc. & WUPS
COCO-QA	COCO	117,684	117,684	1.0	4	Automatic	8.6	1.0	Acc. & WUPS
FM-IQA	COCO	120,360	–	–	–	Human	–	–	Human
VQA-v1-real	COCO	204,721	614,163	3.0	20+	Human	6.2	1.1	Acc. against 10 humans
VQA-v2	COCO	204,721	1,105,904	5.4	–	Human	–	–	Acc. against 10 humans
Visual Genome	COCO	108,000	1,445,322	13.4	7	Human	5.7	1.8	Acc.
Visual7W	COCO	47,300	327,939	6.9	7	Human	6.9	1.1	Acc.

FM-IQA

The FM-IQA dataset [11], which refers to freestyle multilingual image question answering, is also constructed based on the COCO dataset and contains 120,360 images. The most notable difference between FM-IQA and COCO-QA is that the QA pairs are collected by human annotators from the Amazon Mechanical Turk (AMT) crowdsourcing platform. These annotators can specify any type of question related to the given image, thereby increasing the diversity and quality of questions. A total of 250,560 question/answer pairs are collected.

VQA-v1

The VQA-v1 dataset [4] is one of the most widely used VQA datasets constructed based on the COCO dataset, which consists of two parts: VQA-v1-real using natural images and VQA-v1-abstract using synthetic cartoon images. VQA-v1-real contains 123,287 images for training and 81,434 images for testing from the COCO dataset. The question/answer pairs are collected by human annotators, leading to a high diversity, and binary (i.e., yes/no) questions are introduced. Overall, 614,163 questions are collected, each having 10 answers from 10 different annotators. However, this dataset has a large bias, in which several questions can be answered without visual knowledge. For example, for the question starting with "Do you see...", blindly answering "yes" without looking at the image will result in an accuracy of 87%. The aim of the VQA-v1-abstract dataset is to improve the high-level reasoning of VQA models. The VQA-v1-abstract contains 50,000 clipart scenes and a total of 150,000 questions (i.e., three questions per scene), with each question answered by 10 annotators, which are collected in a similar way as in the VQA-v1-real dataset.

VQA-v2

The VQA-v2 dataset is an extensive version of the VQA-v1-real dataset, which aims to solve the large bias in the original dataset. The balanced VQA-v2 dataset is constructed by collecting complementary images that are similar, but the answers of these two images are different. Specifically, for each question, two similar images are collected by AMT workers, and the corresponding answers are different. Overall, the VQA-v2 dataset has 204,721 images and 1,105,904 questions with 10 answers for each question. The number of image-question pairs is twice that in the VQA-v1-real dataset. The balanced VQA-v2 dataset mitigates the biases in the original VQA-v1-real dataset, which prevents VQA models from exploiting language priors to achieve higher evaluated scores and helps develop highly interpretable VQA models that focus more on the visual contents.

Visual Genome

The Visual Genome QA dataset [19] is constructed based on the Visual Genome project [19], which includes unique structured annotations of scene contents in the form of scene graphs . These scene graphs describe the visual elements of the scenes with attributes and the relationships between them. The Visual Genome contains 108,000 images derived from the COCO dataset. The question/answer pairs are collected by human annotators. Two types of questions are considered: free-form and region-based, and the questions must start with "who, what, where, when, why, how or which". In the free-form setting, the annotator is shown an image and asked to provide eight question/answer pairs. To encourage diversity, the annotator is forced to use three different start words among the seven mentioned above. In the region-based setting, the annotator must provide questions/answers related to a specific, given the region of the image. The diversity of the answers in the Visual Genome is larger than that in VQA-real [4], and a key advantage of the Visual Genome dataset for VQA is the potential for using structured scene annotations.

Visual7w

The Visual7w dataset [51] is a subset of Visual Genome that contains 47,300 images and 327,939 questions with additional annotations. The questions are evaluated in a multiple-choice setting in which each question is provided with four candidate answers, among which only one is correct. In addition, all the objects mentioned in the questions are visually grounded, i.e., associated with bounding boxes of their depictions in the images.

4.3 Generation Versus Classification: Two Answering Policies

A classical VQA model usually consists of three components: visual and textual feature extraction from the given image and question, the joint fusion of visual and textual features and answer generation based on the fused features. In general, CNN networks such as VGGNet, ResNet and Faster R-CNN are used to extract image features, and RNN networks such as LSTM and GRU are used to extract question features. For feature fusion, deep learning techniques such as joint embedding and attention mechanisms are utilized. In terms of answer generation, two answering policies exist: question answering as a sequence generation task or as a classification task. As shown in Fig. 4.1, when formulated as a classification task, the joint representation of an image and a question is passed through a neural network classifier, and a single-phrase answer from a predefined vocabulary is produced. In contrast, when formulated as a sequence generation task, a decoder RNN network is used to produce answers with different lengths.

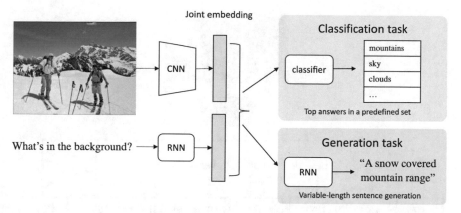

Fig. 4.1 Difference between two polices of generation and classification for VQA tasks

4.4 Joint Embedding Methods

Multimodal joint embedding was first proposed to perform image captioning tasks [9, 21, 42, 44] and has been reinforced in VQA tasks. By projecting both images and questions into a common space, simple and widely used joint embedding methods can be used to accomplish VQA tasks. In this section, we introduce two kinds of joint embedding models: sequence-to-sequence encoder-decoder models and bilinear encoding models.

4.4.1 Sequence-to-Sequence Encoder/Decoder Models

Motivation

With the development of deep learning technology, end-to-end methods have emerged as promising tools for solving computer vision and natural language processing problems. In addition, multimodal learning, such as image captioning, has achieved notable results, based on a multimodal encoder/decoder architecture. Thus, it is intuitive to incorporate the encoder/decoder architecture in VQA methods. To solve challenging VQA tasks, several encoder/decoder VQA models have been proposed, such as neural image-QA and multimodal QA (mQA).

Methods

Malinowski et al. [28] proposed neural image QA, an end-to-end deep learning architecture, to answer natural language questions regarding real-world images in a single holistic and monolithic model. As shown in Fig. 4.2, in neural image QA, a CNN

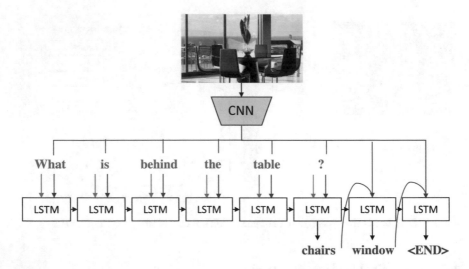

Fig. 4.2 Overview of neural image QA

network extracts image features, and an LSTM network is used to encode questions. These two networks are subsequently combined to generate multiple word answers. In this architecture, answer prediction is formulated as a sequence generation process of multiple words:

$$\hat{a}_t = \operatorname*{argmax}_{a \in \mathcal{V}} p(a|x, q, \hat{A}_{t-1}; \theta), \tag{4.1}$$

where $\hat{A}_{t-1} = \{\hat{a}_1, \ldots, \hat{a}_{t-1}\}$ represents the previous answer words, x and q represent the given image and question, $\mathcal{V}$ represents the answer vocabulary and θ represents the learnable parameters in the model. The given image x is encoded by GoogLeNet, which is pretrained on the ImageNet dataset and fixed except for the last layer. Moreover, the given question q with answer words a are encoded as one-hot vectors and embedded into low-dimensional vectors by a learned embedding network. Subsequently, question q is augmented with answer words a as $\hat{q}$, i.e., $\hat{q} = [q, a]$. Specifically, in the training phase, q is augmented with ground truth answer words a. In the prediction phase, at each time step t, q is augmented with predicted answer words $\hat{a}_{1..t}$ as $\hat{q}_t$, i.e., $\hat{q}_t = [q, \hat{a}_{1..t}]$. Next, the LSTM unit takes v_t as input, which is the concatenation of $[x, \hat{q}_t]$ and predicts the answer word $\hat{a}_t$ at each time step t. In this procedure, the LSTM network predicts a sequence of multiple words as answers until the symbol word <END> is predicted.

Gao et al. [11] proposed a multimodal QA (mQA) model, in which two separate LSTM networks are used to prepare the question and answer. The mQA architecture is different from that of neural image QA, in which the question is augmented with the answer by concatenation and only one LSTM is used. As shown in Fig. 4.3, the mQA consists of four key parts: an LSTM network named LSTM(Q) to extract

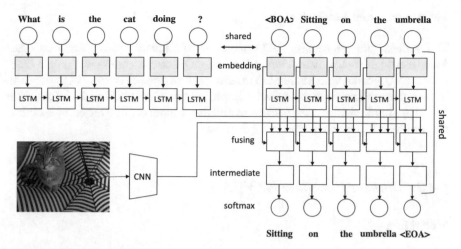

Fig. 4.3 Overview of multimodal QA

the question representation, a CNN network to extract the image features, an LSTM network known as LSTM(A) to extract representations of answer words and a feature fusing network that generates answers. Specifically, GoogLeNet pretrained on the ImageNet dataset is used to extract image features, which are fixed during the QA training procedure. LSTM(Q) and LSTM(A) have similar network structures but do not share the weight parameters, in contrast to the case of neural image QA. The features from the first three components are fused by the last feature fusing component for the tth word as

$$f(t) = g(V_{r_Q} r_Q + V_I I + V_{r_A} r_A(t) + V_w w(t)), \quad (4.2)$$

where "+" represents elementwise addition, r_Q represents the representation of the last word from LSTM(Q), I represents the image feature, $r_A(t)$ represents the hidden representation of LSTM(A) for the tth word, $w(t)$ represents the word embedding of the tth word in the answer, $V.$ represents the learnable weight matrices and $g(.)$ is an elementwise nonlinear function. After fusion, the fused multimodal representation is mapped back to the word representation by an intermediate layer, followed by a fully connected softmax layer to generate answers. In addition, because the same words in both the question and answer must have the same meaning, mQA utilizes a weight-sharing strategy that can reduce parameters and help enhance the performance. In particular, the weight matrices of the word embeddings in both LSTM(Q) and LSTM(A) are shared, and the weight matrix of the word embeddings is shared with the softmax layer in a transposed manner.

Performance and Limitation

The earliest joint embedding VQA methods of sequence-to-sequence encoder/decoder models such as neural image QA and mQA can be considered simple baselines for VQA tasks, which exhibit an inferior performance on several VQA datasets. Indeed, these methods are excessively simple to implement multimodal information fusion by simple elementwise operations and thus cannot capture the complex information embodied in the images and questions.

4.4.2 Bilinear Encoding for VQA

Motivation

Encoding an expressive joint representation of visual and textual features is necessary for VQA systems, which renders it easier to learn a classifier and effectively realize reasoning. In fact, simple methods of joint embedding, such as elementwise products, elementwise sum and concatenation, cannot easily capture the complicated correlations between the visual and textual features. Bilinear pooling models are believed to be more expressive than simple elementwise fusion methods. However, due to their large memory consumption and computational cost, native bilinear pooling models cannot be directly used in VQA tasks. For example, if we set the image and question feature vectors to dimensions of 2048 and incorporate 3000 classes for answers, then the learnable bilinear model will have 12.5 billion parameters. Thus, several multimodal bilinear pooling models that can solve the above problem have been proposed to encode joint representations in VQA tasks, such as MCB and MLB.

Methods

Native bilinear models take the outer product of the visual feature vector x and textual feature vector q as the input and generate a projected feature vector z with a large number of parameters in the learned linear model M as

$$z = M[x \otimes q], \tag{4.3}$$

where $\otimes$ denotes the process of the outer product, and $[\cdot]$ denotes the linearization of a matrix into a vector.

To project the high-dimensional outer product into a lower-dimensional space and indirectly compute the outer product, Fukui et al. [10] proposed multimodal compact bilinear pooling (MCB), which utilizes a count sketch projection function Ψ [6] and projects a vector $v \in \mathbb{R}^n$ to a vector $y \in \mathbb{R}^d$. This count sketch procedure is completed using two vectors $s \in \{-1, 1\}^n$ and $h \in \{1, ..., d\}^n$, which are initialized randomly from a uniform distribution and fixed during future invocations of the count sketch. In addition, y is initialized to a zero vector. Specifically, s is used to map the value

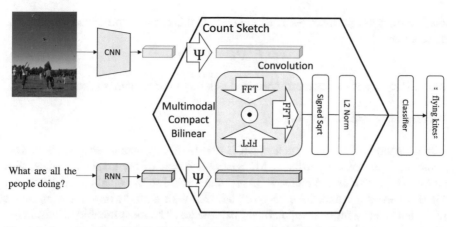

Fig. 4.4 Overview of multimodal compact bilinear pooling for VQA

v_i of each element in the input vector v to value v_i or $-v_i$, while h is used to map each index i in the input vector v to the index j in the output vector y. For each element v_i in input vector v, the destination index is computed as $j = h_i$, and the corresponding value is obtained as $y_j = s_i \cdot v_i$. Through this procedure, the outer product can be projected to a lower dimensional space, thereby reducing the large numbers of parameters in W. As shown in Fig. 4.4, both visual vector x and textual vector q are projected to count sketch vectors of x' and q' using Ψ. Furthermore, to indirectly and effectively compute the outer product, the count sketch of the outer product of x and q can be computed as follows:

$$\Psi \left(x \otimes q, h, s \right) = \Psi \left(x, h, s \right) * \Psi \left(q, h, s \right) = x' * q', \qquad (4.4)$$

where $*$ represents the convolution process. According to the convolution theorem, the convolution of $x' * q'$ in the time domain can be rewritten in the frequency domain as follows:

$$x' * q' = FFT^{-1} \left(FFT \left(x' \right) \odot FFT \left(q' \right) \right), \qquad (4.5)$$

where $\odot$ represents the elementwise product, $FFT(\cdot)$ represents the fast Fourier transform and $FFT^{-1}(\cdot)$ represents the inverse fast Fourier transform. At present, the procedure of multimodal compact bilinear pooling has been established, and this joint representation is considerably more expressive than those derived from simple joint embedding methods.

Although the MCB involves considerably fewer parameters than native bilinear models, it still generates high-dimensional features and is computationally complex. To further reduce the number of parameters, Kim et al. [17] proposed multimodal low-rank bilinear pooling (MLB). The procedure of generating projected vector z by bilinear models can be rewritten as $z = x^T W q$, where W is a high-rank weight matrix. The core idea of MLB is to factor the large weight matrix W into two small and low-rank weight matrices as $W = UV^T$. In this case, the projected vector z can be represented as

$$z = P^T(U^T x \circ V^T q), \tag{4.6}$$

where P represents a matrix of ones, and ∘ denotes the Hadamard product.

Performance and Limitation

Bilinear encoding methods such as MCB and MLB have achieved notable performance in VQA tasks. In particular, MCB exhibited state-of-the-art performance, with an overall score of 66.5% on the VQA-v1-real test-std set for open-ended questions. MLB achieved a competitive score of 66.89% with a slight improvement, albeit with significantly fewer computational parameters. The most notable disadvantage of bilinear encoding methods is the large computation cost.

4.5 Attention Mechanisms

Attention mechanisms have been widely and effectively used in computer vision and natural language processing tasks. It is intuitive to utilize the attention mechanism in VQA tasks, and the performance of attention-based methods is promising. In this section, we describe several classic attention-based VQA models, such as stacked attention networks (SANs), hierarchical question-image co-attention (HieCoAtt), and bottom-up and top-down attention (BUTD).

4.5.1 Stacked Attention Networks

Motivation

A common practice in VQA tasks is to use a CNN network to extract global image features and an RNN network to extract holistic question features. However, this simple process cannot address complicated VQA tasks, which often require multiple-step fine-grained reasoning. The global image feature is expected to introduce noise of irrelevant image regions in the VQA models. Moreover, a single-step attention mechanism is not adequately effective to identify the correct region in the case of a complicated question. Consequently, Yang et al. [43] proposed a stacked attention network (SAN) to implement multiple-step fine-grained reasoning for VQA tasks, based on a multi-layer attention mechanism.

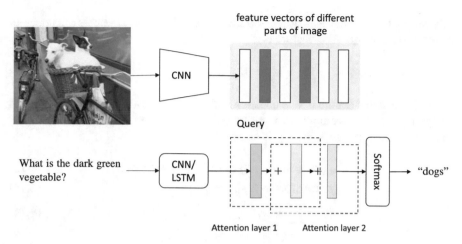

feature vectors of different
parts of image

Query

What is the dark green
vegetable?

CNN/
LSTM

Softmax

"dogs"

Attention layer 1   Attention layer 2

Fig. 4.5 Overview of the SAN

Method

As shown in Fig. 4.5, the SAN consists of three major parts: image feature extraction, question feature extraction and stacked attention.

For image feature extraction, SAN utilizes the last pooling layer of VGGNet to extract image feature f_I, which can preserve the spatial information of input image I:

$$f_I = \text{CNN}_{vgg}(I). \tag{4.7}$$

The image feature f_I has dimensions of $512 \times 14 \times 14$, where 14×14 represents the number of regions in the input image I. Subsequently, each feature vector f_i in all 196 regions is transformed into the final image feature V_I that shares the same dimension as that of the question vector V_Q:

$$v_I = \tanh(W_I f_I + b_I). \tag{4.8}$$

Given the one-hot representation of T question words $q = [q_1, ..., q_T]$, SAN uses two methods to extract the question features: an LSTM-based method and a CNN-based method. In the LSTM-based method, question q is embedded into vectors and fed into an LSTM network, where the hidden state h_T of the last layer is considered as question feature v_Q. In the CNN-based method, the SAN first embeds a one-hot word representation to vectors $x = [x_1, ..., x_T]$. Subsequently, SAN uses multiple convolution kernels and max pooling to generate unigram, bigram and trigram text features $\tilde{h}_1, \tilde{h}_2, \tilde{h}_3$ as

$$h_{c,t} = \tanh(W_c x_{t:t+c-1} + b_c), \tag{4.9}$$

$$\tilde{h}_c = \max_t [h_{c,1}, h_{c,2}, ..., h_{c,T-c+1}], \tag{4.10}$$

where $c = 1, 2, 3$ represents different kernel sizes. Next, these features are concatenated as the final question feature v_Q:

$$v_Q = [\tilde{h}_1, \tilde{h}_2, \tilde{h}_3]. \tag{4.11}$$

With the image feature v_I and question feature v_Q, attention weights p_I over each image region vi are computed for v_Q by using a one-layer network with softmax, and the attended image feature $\tilde{v}_I$ is obtained:

$$h_A = \tanh(W_{I,A}v_I \oplus (W_{Q,A}v_Q + b_A)), \tag{4.12}$$

$$p_I = \text{softmax}(W_P h_A + b_P), \tag{4.13}$$

$$\tilde{v}_I = \sum_i p_i v_i, \tag{4.14}$$

where $W.$ represents learnable weights, $b.$ represents biases and $\oplus$ represents the addition of a matrix and a vector.

Next, $\tilde{v}_I$ is combined with V_Q as a query vector u for the multiple attention process:

$$u = \tilde{v}_I + v_Q. \tag{4.15}$$

Specifically, the SAN uses multiple attention layers, and for the kth attention layer, query vector u_{k-1} is used to generate the attended image feature $\tilde{v}_I^k$. Next, a new query vector, u^k, is obtained by summing u_{k-1} and $\tilde{v}_I^k$, and this process is repeated K times until the final query vector u^K is produced:

$$h_A^k = \tanh(W_{I,A}^k v_I \oplus (W_{Q,A}^k u^{k-1} + b_A^k)), \tag{4.16}$$

$$p_I^k = \text{softmax}(W_P^k h_A^k + b_P^k), \tag{4.17}$$

$$\tilde{v}_I^k = \sum_i p_i^k v_i, \tag{4.18}$$

$$u^k = \tilde{v}_I^k + u^{k-1}, \tag{4.19}$$

where u^0 is set as v_Q.

Finally, u^K is used to predict the answer:

$$p_{\text{ans}} = \text{softmax}(W_u u^K + b_u). \tag{4.20}$$

Performance and Limitation

The stacked attention network (SAN) achieves an overall score of 58.9% on the VQA-v1 test-std set, thereby outperforming the best VQA baseline by 4.8% and existing state-of-the-art methods by a considerable margin on the two VQA datasets of DAQUAR and COCO-QA. An ablation study shows that two-layer SANs outperform one-layer SANs, which demonstrates the positive impact of using multiple attention layers.

4.5.2 Hierarchical Question-Image Co-Attention

Motivation

The existing attention-based approaches for VQA tasks, in which only question-guided visual attention is implemented, focus is only on where to look. However, it is as important to know where to listen. Furthermore, image-guided question attention can mitigate the linguistic noise of variable questions in VQA tasks. Considering this aspect, Lu et al. [23] proposed hierarchical question-image co-attention (HieCoAtt) to address VQA tasks, which can jointly implement co-attention with both question-guided image attention and image-guided question attention.

Method

To implement hierarchical question-image co-attention, two novel components are proposed: question hierarchy and co-attention.

Given the one-hot representation of T question words, $Q = \{q_1, \ldots, q_T\}$, the question hierarchy module generates three-level representations: word-level embedding q_t^w, phrase-level embedding q_t^p and question-level embedding q_t^s for each position t. Specifically, one-hot question words are embedded in the word-level representation as $Q^w = \{q_1^w, \ldots, q_T^w\}$. The phrase-level representation is obtained by convolution with multiple kernels and max pooling, which is similar to the CNN-based question feature extraction in the SAN. The generation of the phrase-level representation can be expressed as follows:

$$\hat{q}_{s,t}^p = \tanh(W_c^s q_{t:t+s-1}^w), \quad s \in \{1, 2, 3\}, \tag{4.21}$$

$$q_t^p = \max(\hat{q}_{1,t}^p, \hat{q}_{2,t}^p, \hat{q}_{3,t}^p), \quad t \in \{1, 2, \ldots, T\}, \tag{4.22}$$

where s represents different windows sizes. Question-level representation q_t^s is the hidden state of the LSTM network at time t, where phrase-level representation q_t^p is encoded.

Two co-attention mechanisms are proposed, namely, parallel co-attention and alternating co-attention. We present details of the more representative parallel co-attention.

As shown in Fig. 4.6, in parallel co-attention, the co-attention is simultaneously implemented between images and questions. First, given the feature map V and corresponding question Q, a similarity matrix C is computed as

$$C = \tanh(Q^T W_b V). \tag{4.23}$$

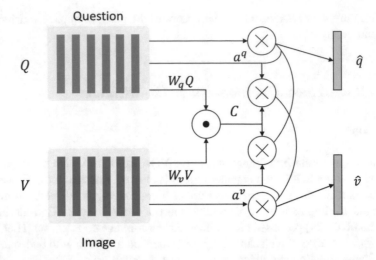

Fig. 4.6 Illustration of parallel co-attention in the hierarchical question-image co-attention model

Using C, the attention score a^v for each location of the image and a^q for each location of the question are simultaneously computed:

$$H^v = \tanh(W_v V + (W_q Q)C), \quad H^q = \tanh(W_q Q + (W_v V)C^T)$$
$$a^v = \text{softmax}(w_{hv}^T H^v), \quad a^q = \text{softmax}(w_{hq}^T H^q), \tag{4.24}$$

where W. and w. denote the learnable weight matrices and vectors, respectively. The attended image feature $\hat{v}$ and question feature $\hat{q}$ are obtained:

$$\hat{v} = \sum_{n=1}^{N} a_n^v v_n, \quad \hat{q} = \sum_{t=1}^{T} a_t^q q_t. \tag{4.25}$$

In alternating co-attention, co-attention operates in a sequential manner. Specifically, the model first generates attended image features under the guidance of the question feature and later attends to the question feature under the guidance of the attended image feature.

Both these co-attention mechanisms are implemented in the hierarchy architecture, which generates hierarchical attended features $\hat{v}^r$ and $\hat{q}^r$, where $r \in \{w, p, s\}$.

Performance and Limitation

The hierarchical question-image co-attention (HieCoAtt) method achieves an overall score of 62.1% for open-ended questions and 66.1% for multiple-choice questions on the VQA-v1 test-std set, thereby outperforming other state-of-the-art methods with a margin of at least 1.7%. Qualitative results show that the hierarchical architecture in the proposed co-attention can capture complementary information well from each level, which can help understand both questions and images. However, the parallel co-attention is more difficult to train, whereas the alternating co-attention may suffer from accumulated errors.

4.5.3   Bottom-Up and Top-Down Attention

Motivation

Attention mechanisms have been widely used in VQA tasks and proven to be effective. These attention-based methods often operate in a top-down and task-specific manner, thereby computing a soft attention score over each grid region of the image under the guidance of the question, treating all grid regions equally. This framework is similar to a human vision system, in which humans focus on a specific region according to the task context, such as searching for something. In addition to top-down attention, there exists a bottom-up attention mechanism in the human vision system. Specifically, humans are automatically attracted by salient objects or scenes. Salient regions in an image are considerably more expressive than grid regions and must be focused on. Thus, Anderson et al. [1] proposed a combined bottom-up and top-down attention model (BUTD) for VQA tasks. In this framework, bottom-up attention is implemented by detecting salient regions, and top-down attention is implemented by computing attention scores over the proposed regions according to the question context.

Method

As shown in Fig. 4.7, given an image I, the BUTD first uses the Faster R-CNN network to propose the top K salient regions. These K salient regions are passed through the ResNet-101 network to generate image features $V = \{v_1, ..., v_K\}$, where v_i is a 2048-D feature vector, which represents the visual feature for each salient region. K can be either a fixed value $K = 36$ or an adaptive value from 1 to 100. Both Faster R-CNN and ResNet-101 are pretrained on the ImageNet dataset and trained on the Visual Genome dataset. In addition, the BUTD involves an extra

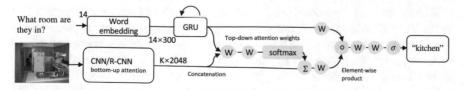

Fig. 4.7 Overview of the bottom-up and top-down attention model for VQA

output for Faster R-CNN, which predicts the attributes of the detected regions to enhance the performance.

Using the bottom-up attention and generated image features V, BUTD implements soft top-down attention under the guidance of the given question. The question is first trimmed to a fixed length with 14 maximum words and embedded in vectors with dimensions of 14×300 initialized by GloVe. Subsequently, a GRU network is used to extract question embedding q, which is the last hidden state of the GRU network with dimensions of 512. Each visual feature v_i of the proposed regions is concatenated with q and passed to a nonlinear layer f_a with softmax to compute the attention score:

$$a_i = w_a f_a([v_i, q]), \tag{4.26}$$
$$\alpha = softmax(a), \tag{4.27}$$

where $[\cdot]$ represents concatenation, and w_a is a learnable weight. With the soft attention score over each salient region, the attended image feature $\hat{v}$ is generated as follows:

$$\hat{v} = \sum_{i=1}^{K} \alpha_i v_i. \tag{4.28}$$

With both question embedding q and attended image feature $\hat{v}$, the joint representation h of the image and question is obtained as follows:

$$h = f_q(q) \circ f_v(\hat{v}), \tag{4.29}$$

where $\circ$ represents the Hadamard product. Subsequently, the classification score $\hat{s}$ for each candidate answer is computed:

$$\hat{s} = \sigma(W_o f_o(h)), \tag{4.30}$$

where σ represents the activation functions, and $W - o$ is a learnable weight matrix.

Performance and Limitation

The bottom-up and top-down attention method (BUTD) achieves an overall score of 70.34% with 30 ensembled models on the VQA-v2 test-std set and ranked first in the VQA Challenge 2017, outperforming other state-of-the-art methods with a large margin across all question types. In addition, this framework has been the most widely used baseline for VQA research since its realization. However, the capacity of the Faster R-CNN object detector influences the performance of the VQA models, and an increased capacity can help extract more expressive detection features.

4.6 Memory Networks for VQA

Memory networks have been noted to be effective tools for question answering tasks in NLP, which can explore fine-grained features with previous interactions. Thus, it is natural to utilize these memory networks in VQA tasks. In this section, we describe two classic memory networks for VQA tasks: improved dynamic memory networks (DMNs+) and memory-augmented networks (MANs).

4.6.1 Improved Dynamic Memory Networks

Motivation

The existing work on dynamic memory networks (DMNs) has demonstrated their considerable potential in accomplishing natural language processing tasks, especially question answering (QA). However, this method requires additional labeled supporting facts and is difficult to apply to other modalities. Thus, Xiong et al. [39] proposed improved dynamic memory networks (DMNs+) for VQA tasks, which can directly manage image data.

Method

The DMN for question answering consists of four main modules: An input module to process the input text data as facts F, a question module to embed a question as feature vector q, an episodic memory module to retrieve required information from facts F and an answer module to predict answers. To adapt the DMN to VQA tasks, DMN+ modifies the input module and episodic memory module.

As shown in Fig. 4.8, in the input module, DMN+ uses a VGG network to extract 196 local region features with dimensions of 512 and a linear network to project these feature vectors into the same space with question feature vector q. Subsequently, these

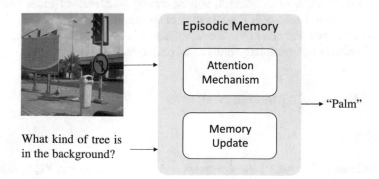

Fig. 4.8 Illustration of improved dynamic memory networks for VQA

local feature vectors are passed through a bi-directional GRU network to generate globally aware feature vectors known as "facts" $\overleftrightarrow{F} = [\overleftrightarrow{f_1}, \ldots, \overleftrightarrow{f_N}]$ as inputs of the episodic memory module.

The episodic memory module retrieves information from the facts that are needed to answer the question. Specifically, the module includes an attention mechanism for selecting relevant facts that allows interactions between facts, questions and previous memory states and a memory update mechanism that generates a new memory representation through the interaction between the current state and retrieved facts. The attention mechanism is implemented using an attention-based GRU network to generate the contextual vector c_t for the update of the episodic memory state m_t.

$$z_i^t = [\overleftrightarrow{f_i} \circ q; \overleftrightarrow{f_i} \circ m^{t-1}; |\overleftrightarrow{f_i} - q|; |\overleftrightarrow{f_i} - m^{t-1}|], \tag{4.31}$$

$$Z_i^t = W^{(2)} \tanh\left(W^{(1)} z_i^t + b^{(1)}\right) + b^{(2)}, \tag{4.32}$$

$$g_i^t = \frac{\exp(Z_i^t)}{\sum_{k=1}^{M_i} \exp(Z_k^t)}, \tag{4.33}$$

$$h_i = g_i^t \circ \tilde{h}_i + (1 - g_i^t) \circ h_{i-1}, \tag{4.34}$$

where $\overleftrightarrow{f_i}$ is the ith fact, q is the question vector, m^{t-1} is the state of the previous episode memory, h is the hidden state of the GRU network, $\circ$ represents the elementwise product, $|\cdot|$ represents the elementwise absolute function and $[;\,]$ represents concatenation. The contextual vector c_t is the last hidden state of the GRU network, and the memory update is implemented by a ReLU layer:

$$m^t = ReLU\left(W^t[m^{t-1}; c^t; q] + b\right). \tag{4.35}$$

Finally, the answer module uses the final state of the memory network and question vector to predict the output of a single word or multiple word sentence.

Performance and Limitation

The improved dynamic memory network (DMN+) method achieves an overall score of 60.4% on the VQA-v1 test-std set, outperforming other state-of-the-art methods with a margin of at least 1.5%. For all types of questions, DMN+ achieves state-of-the-art performance. For other questions, the margin is up to 1.8%. However, DMN+ cannot effectively address a large number of problems.

4.6.2 Memory-Augmented Networks

Motivation

The distribution of natural language question/answer pairs in the VQA dataset is often heavy-tailed, and VQA models tend to respond to the majority of training data, neglecting specific scarce but important exemplars. A common practice is to mark

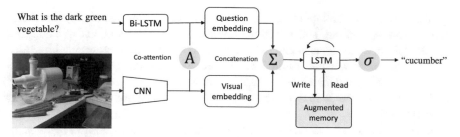

Fig. 4.9 Overview of memory-augmented networks

the rare words in questions as unknown tokens and exclude rare answers directly in training data. In addition, VQA models are also inclined to learn simply from question/answer pairs without understanding visual contents, known as the language bias problem in VQA. To solve the above heavy-tailed problem and language bias problem, Ma et al. [24] proposed memory-augmented networks (MANs) inspired by memory-augmented neural networks and co-attention mechanisms.

Method

MAN utilizes co-attention to jointly embed image and question features with a followed memory-augmented network to remember scarce exemplars in training data. The augmented-memory network in MAN contains both an internal memory inside LSTM and an external memory controlled by LSTM, and this framework is considerably different from that of the DMN. As shown in Fig. 4.9, the MAN consists of four components: input module, sequence co-attention module, memory-augmented network and answer reasoning module.

For the image input, the MAN utilizes pretrained VGGNet-16 and ResNet-101 to extract image features $\{v_n\}$ with spatial layout information, which are derived from the outputs of the last pooling layer, corresponding to 14×14 spatially distributed regions. For the question input, the embedded word tokens w_t are fed into bidirectional LSTMs to generate fixed-length sequential word vectors as question features $\{q_t\}$:

$$h_t^+ = \mathrm{LSTM}(w_t, h_{t-1}^+), \tag{4.36}$$

$$h_t^- = \mathrm{LSTM}(w_t, h_{t+1}^-), \tag{4.37}$$

$$q_t = [h_t^+, h_t^-], \tag{4.38}$$

where h_t^+ and h_t^- represent the hidden states of the forward and backward LSTMs at time step t, and [.] represents concatenation.

Subsequently, given the image and question features, a sequential co-attention mechanism is utilized to attend to the most relevant parts of features for each modality according to the other modality. First, $\{v_n\}$ and $\{q_t\}$ are summed and averaged to feature vectors v_0 and q_0, respectively. In addition, an elementwise product is implemented on v_0 and q_0 to generate the joint base vector m_0. Using visual feature vectors v_n and m_0, soft attention weights α_n are computed, and the attended visual feature vector v^* is generated using a two-layer neural network with a softmax layer. Similarly, using question feature vectors a_t and m_0, soft attention weights α_t are computed, and the attended visual feature vector q^* is generated. Finally, v^* and q^* are concatenated to represent the co-attended image and question features as $x_t = [v_t^*, q_t^*]$.

Considering the concatenated attended visual features and question features x_t, the MAN adopts a memory-augmented neural network to enhance the effect of scarce training data during the training process. The memory-augmented neural network uses an LSTM controller consisting of an internal memory that receives input data and an external memory M_t from which and to which external information is read and written. The feature vector x_t is first passed to the LSTM controller, and the hidden state h_t is obtained, which is considered the query for M_t. Subsequently, the cosine distance $D(h_t, M_t(i))$ between h_t and each element $M_t(i)$ in the external memory is computed and normalized as the attention weight $w_t^r(i)$ for the read process by softmax. Using these read weights, the attended read memory r_t is generated:

$$h_t = \textbf{LSTM}(x_t, h_{t-1}), \tag{4.39}$$

$$D(h_t, M_t(i)) = \frac{h_t \cdot M_t(i)}{\|h_t\|\|M_t(i)\|}, \tag{4.40}$$

$$w_t^r(i) = \text{softmax}\big(D(h_t, M_t(i))\big), \tag{4.41}$$

$$r_t = \sum_i w_t^r(i)M_i. \tag{4.42}$$

Finally, r_t is concatenated with h_t to generate the final feature vector o_t as the input for the answer classifier. Specifically, the answer classifier consists of a one-layer perceptron with a softmax function:

$$h_t = \tanh(bW_o o_t), \tag{4.43}$$

$$p_t = \text{softmax}(W_h h_t). \tag{4.44}$$

Performance and Limitation

The memory-augmented networks (MAN) method achieves a competitive performance compared to the state-of-the-art method of MCB on both the VQA-v1 and VQA-v2 test sets. On the VQA-v1 dataset, the MAN exhibits a slightly enhanced performed on multiple-choice questions and slightly deteriorated performance on open-ended questions. Compared to DMN+, which exploits only the internal memory inside RNNs rather than an augmented external memory, the MAN exhibits a higher performance with a large margin of 3.5%. On the VQA-v2 test set, the performance of the MAN is slightly decreased by approximately 0.2% compared to that of the MCB.

4.7 Compositional Reasoning for VQA

VQA models are required to implement complex reasoning, which is difficult for a single holistic model to manage.

Modular methods are emerging effective tools for compositional reasoning in VQA tasks, which connect different modules designed for different functions. Specifically, modular networks decompose a question into several components and assemble different networks to predict the answer. In this section, we mainly discuss two compositional reasoning models, namely, the neural module network (NMN) and dynamic neural module network (D-NMN).

4.7.1 Neural Modular Networks

Motivation

In VQA tasks, the questions are often complicated and multiple processing steps are required to identify the correct answer. For example, given the simple question of *"What color is the dog?"*, the VQA models must first locate the dog and then recognize the color of the dog. However, even with advanced deep learning methods, it is difficult for a single optimal network to manage all subtasks. Thus, Andreas et al. [3] proposed neural module networks (NMNs), which decompose the problem into a multistep process by using compound modular networks to predict the final answer.

Method

As shown in Fig. 4.10, the NMN consists of a set of modular networks (modules), which are assembled by a network layout predictor. Especially for VQA tasks, the NMN adds an LSTM question encoder to provide underlying syntactic and semantic knowledge.

The NMN includes five modules: find module, transform module, combine module, describe module and measure module. These five modules are utilized to address three types of data: images, unnormalized attentions and labels. Specifically, as shown in Fig. 4.11, the find module implements unnormalized attention over all regions of the input image by convolution layers such as *find[cat]*. The transform module refines one attention to another attention by using a multi-layer perceptron, which shifts the input attention to other required regions such as the *transform[above]*. The combine module fuses two attentions into one attention, for instance, through *combine[and]*, which only activates the intersection regions of the two attentions using a convolution layer with ReLU. The describe module takes the given image and attention as the input and predicts a distribution over labels to problems except *yes/no* problems such as *describe[where]*. The measure module is similar to the describe module; however, it only predicts the distribution over labels to the *yes/no* problem such as

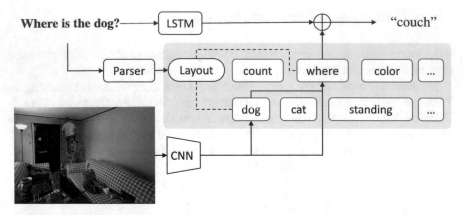

Fig. 4.10 Overview of the neural module network for VQA

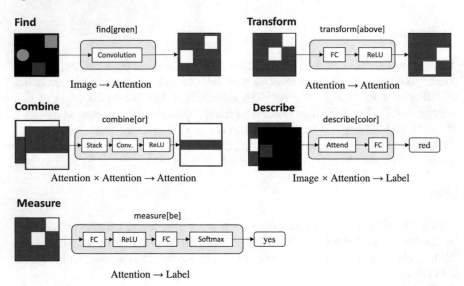

Fig. 4.11 Modules of the neural module network

measure[is]. Notably, these modules are trained together in the assembled model rather than in isolation.

Using the abovementioned modular networks, the NMN generates layouts of the required networks and assembles these networks according to the given question. Specifically, the NMN uses a Stanford parser [8] to generate the filtered dependency representation. For example, the question *"what color is the cat?"* is transformed to *color(cat)*. Subsequently, using the dependency representation, the layouts are generated according to the rules as follows: leaf nodes that take the image as input are implemented through the find module; intermediate nodes are implemented by the transform module or combine module; and root nodes compute the final output and are implemented by the describe module or measure module. After layout generation, the question *"what color is the cat?"* is transformed to *describe[color](find[cat])*.

Finally, the root representation of the assembled modular networks is summed up with the last hidden state of an LSTM question encoder to predict the final answer with a fully connected layer and softmax.

Performance and Limitation

The neural module network (NMN) method achieves an overall score of 58.0% on the VQA-v1 test-dev set, outperforming other state-of-the-art methods. NMN performs especially well on questions answered in terms of an object, attribute or number. However, the use of a superior parser or joint learning can help reduce parser errors, thereby enhancing the performance of VQA tasks.

4.7.2 Dynamic Neural Module Networks

Motivation

The existing frameworks of NMN use manually specified modular structures, which are chosen by the syntactic processing of questions. These handwritten rules to deterministically transform dependency trees into layouts limit the models' capacity to produce complicated structures in which large variations in the network structure per question are not permitted. To solve more difficult problems that require enhanced generalization of more structured semantic representations, Andreas et al. [2] proposed a dynamic neural module network (D-NMN), which extends the NMN's mechanism of decomposing VQA tasks into a sequence of modular subproblems. The D-NMN can automatically learn module layouts from a set of generated candidates with a structure predictor. In addition, D-NMN can reason over structured information such as knowledge bases in addition to unstructured information such as images.

Method

As shown in Fig. 4.12, the D-NMN consists of two parts: a layout model that automatically chooses module layouts according to the given question and an execution model that predicts answers according to the layouts and world representations (images or knowledge bases). Given question x, world representations w, and collection of model parameters θ, these two models compute two distributions $p(z|x; \theta_l)$ and $p_z(y|w; \theta_e)$, respectively, where z represents the network layout, and y represents the answer.

D-NMN first utilizes a fixed syntactic parser (Stanford parser) to generate a small set of layout candidates, similar to the process of building module layouts in the NMN. With these candidate layouts, the D-NMN uses neural networks with the MLP to rank the candidates. Specifically, an LSTM encoding representation $h_q(x)$

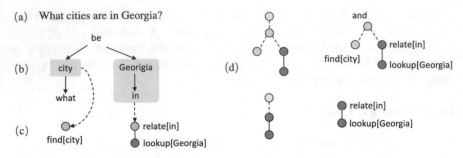

Fig. 4.12 Generation of layouts in a dynamic neural module network

of question q and feature vector representation $f_z(i)$ of layout z_i are passed to an MLP neural network, resulting in a score $s\,(z_i|x)$ for layout z_i:

$$s(z_i|x) = a^\top \sigma(Bh_q(x) + Cf(z_i) + d),\qquad\qquad(4.45)$$

where a, B, C and d are learnable parameters. The scores are normalized using softmax to obtain the distribution $p\,(z_i|x, \theta_l)$, which is used to select the best module layout:

$$p(z_i|x; \theta_l) = \frac{e^{s(z_i|x)}}{\sum_{j=1}^{n} e^{s(z_j|x)}}.\qquad\qquad(4.46)$$

When the module layout z is selected, the execution model assembles the corresponding modules with the world representations into a full neural network. Subsequently, the answer distribution $p_z\,(y|w, \theta_e)$ is obtained according to the intermediate results flowing between compositional modules. The following modules are used in the D-NMN: A lookup module that manages proper nouns and produces one-hot attention over the input feature map; a find module that manages ordinary nouns with verbs and produces an attention over each position of the input feature map; a relate module that addresses prepositional phrases and produces an attention from one region to another; an and module that produces an intersection of attentions and joins layout fragments; a describe module that predicts answers according to the input attention; and an exist module that predicts existential answers according to the input attention. Furthermore, the same module shares the same parameters for different instances.

In addition, the D-NMN utilizes a policy gradient method to transform the non-differentiable selection of z into a differentiable process. Thus, the layout model and the execution model can be trained jointly, in which the parameters of both the layout predictor and modules are simultaneously learned.

Performance and Limitation

The dynamic neural module network (D-NMN) method achieves an overall score of 58.0% on the VQA-v1 test-std set, with high explainability. However, the D-NMN must carefully design submodules. These predefined modules cannot be extended to different datasets. Thus, the feasibility of neural module networks remains a challenge.

4.8 Graph Neural Networks for VQA

The existing CNN-based methods for VQA tasks cannot effectively model the relationships between salient objects in the given image. In addition, these methods lack adequate interpretability for the model performance. Graph learning can effectively address the abovementioned two problems. Thus, it is natural to utilize graph neural networks in VQA tasks. In this section, we present a detailed description of graph convolutional networks (GCNs), graph attention networks (GATs) and graph convolution networks for VQA (graph learners).

4.8.1 Graph Convolutional Networks

Motivation

A mass of real-world data can be represented as a graph, a data structure that models objects and their relationships using nodes and edges, such as social networks, traffic networks and knowledge bases. Convolutional neural networks (CNNs) applied to nonstructural and Euclidean data such as images and texts cannot effectively address such structural and non-Euclidean data of graphs. As shown in Fig. 4.13, the two kinds of data exhibit notable differences. To address graph data well, Kipf and Welling [18] proposed graph convolutional networks (GCNs), which can learn features automatically from objects (nodes) and their relationships (edges). With a promising performance and high interpretability, GCNs are being widely used in solving graph-based problems.

Method

As shown in Fig. 4.14, the propagation of the l-th layer in a neural network can be formulated as follows:

$$H^{l+1} = f\left(H^l, W^l\right), \tag{4.47}$$

where H^l is the feature representation of the lth layer, $f\left(\cdot\right)$ is the function of propagation and W^l is a learnable weight matrix for the lth layer.

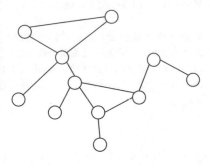

Fig. 4.13 Difference between image data and graph data

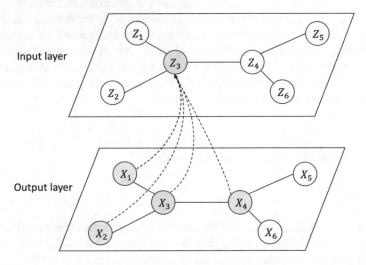

Fig. 4.14 Illustration of the propagation of graph convolutional networks

For example, a simple propagation function in CNNs can be written as

$$f\left(H^l, W^l\right) = \sigma\left(H^l W^l\right),\qquad(4.48)$$

where σ represents nonlinear activation functions such as ReLU, and the bias is omitted for simplification.

In contrast to CNNs, GCNs operate on graphs, and the propagation involves structural information. Specifically, a graph is defined as follows: $\mathcal{G} = (\mathcal{V}, \mathcal{E})$, where $\mathcal{V}$ is the set of nodes that represent objects, and $\mathcal{E}$ is the set of edges that represent the relationships between the objects. GCNs take two matrices as the input: a matrix of node features $X \in \mathbb{R}^{N \times F}$, where N is the number of nodes and F is the input feature dimension of each node, and an adjacency matrix A, the element A_{ij} of which is 1 if nodes i and j are connected. The target of GCNs is to output a transformed feature matrix $Z \in \mathbb{R}^{N \times F'}$, where F' is the output feature dimension of each node. Thus, a simple propagation function of GCNs can be expressed as

$$f\left(H^l, W^l\right) = f\left(H^l, A\right) = \sigma\left(A H^l W^l\right) \tag{4.49}$$

with $H^0 = X$ and $H^L = Z$, where L is the number of layers in the GCNs.

However, this propagation function, in which the features of nodes H^l are simply multiplied with the adjacency matrix A, has two shortcomings: (**i**) For each node i in the graph, this function takes into account features from all neighboring nodes but neglects its own features; (**ii**) High-degree nodes have large values in their transformed features, which may result in vanishing or exploding gradient and high sensitivity of the model to the data scale.

To solve these two problems, GCNs first insert self-loops into each node by adding the identity matrix I to A as

$$\hat{A} = A + I. \tag{4.50}$$

Second, inspired by the common practice in matrix normalization, specifically, the multiplication of the matrix by a diagonal matrix, GCNs use the degree matrix $\hat{D}$ of $\hat{A}$ to normalize $\hat{A}$ in a symmetric manner:

$$\tilde{A} = \hat{D}^{-\frac{1}{2}} \hat{A} \hat{D}^{-\frac{1}{2}}. \tag{4.51}$$

Subsequently, the final propagation in GCNs can be formulated as follows:

$$f\left(H^l, A\right) = \sigma\left(\tilde{A} H^l W^l\right) = \sigma\left(\hat{D}^{-\frac{1}{2}} \hat{A} \hat{D}^{-\frac{1}{2}} H^l W^l\right). \tag{4.52}$$

4.8.2   Graph Attention Networks

Motivation

Although GCNs exhibit a promising performance on graph-structured data, GCNs are structure-dependent and transductive. In other words, a GCN network trained on one graph is difficult to generalize to another graph with a different structure. The attention mechanism can effectively address this shortcoming by specifying different importance scores to different neighboring nodes rather than treating all neighboring nodes equally, as GCNs. Thus, Velickovic et al. [35] proposed graph attention networks (GATs) to manage graph-structured data, which incorporate a self-attention mechanism with no requirement of knowing the structure of the graph in advance and can be easily transferred to other structures of graphs. In other words, GATs are structure-independent and inductive.

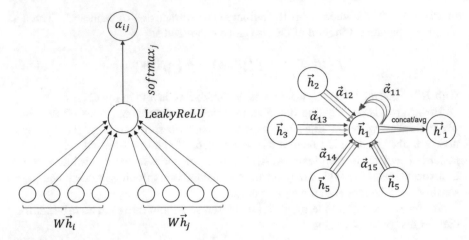

Fig. 4.15 Illustration of graph attention networks

Method

GATs take a set of node features $\mathbf{h} = \{h_1, h_2, \ldots, h_N\}, h_i \in \mathbb{R}^F$ as the input and aim at generating a new set of transformed node features $\mathbf{h}' = \{h'_1, h'_2, \ldots, h'_N\}, h'_i \in \mathbb{R}^{F'}$, where N is the number of nodes, and F and F' are dimensions of the input and output nodes.

As shown in Fig. 4.15, GATs first transform each input node feature vector h_i to a higher-level and more expressive feature vector as $\mathbf{W}h_i$ by using a linear transformation matrix $\mathbf{W} \in \mathbb{R}^{F' \times F}$. For each node i, a pairwise attention coefficient is computed through additive self-attention between every neighboring node j connected to node i (including node i itself) as

$$e_{ij} = \text{LeakyReLU}\left(\mathbf{a}^T \left[\mathbf{W}h_i \parallel \mathbf{W}h_i\right)\right]), \tag{4.53}$$

where $\mathbf{a}$ is a learnable weight vector, $\cdot^T$ is the transposition process and $\parallel$ is the concatenation process.

Next, the attention coefficients are normalized using the softmax function:

$$\alpha_{ij} = \frac{\exp\left(\text{LeakyReLU}\left(\mathbf{a}^T [\mathbf{W}h_i \parallel \mathbf{W}h_j]\right)\right)}{\sum_{k \in \mathcal{N}_i} \exp\left(\text{LeakyReLU}\left(\mathbf{a}^T [\mathbf{W}h_i \parallel \mathbf{W}h_k]\right)\right)}, \tag{4.54}$$

where $\mathcal{N}_i$ represents the set of neighboring nodes of node i.

With the normalized attention coefficients, the final feature propagation of each node can be formulated as

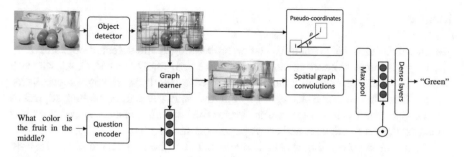

Fig. 4.16 Overview of graph convolutional networks of the proposed graph learner for VQA

$$h_i' = \sigma \left(\sum_{j \in \mathcal{N}_i} \alpha_{ij} \mathbf{W} h_j \right). \tag{4.55}$$

In addition, to enhance the learning capacity and stability, GATs utilize *multihead attention*. Specifically, in the intermediate layers of GATs, K independent attention mechanisms are used, and their corresponding transformed features are concatenated as

$$h_i' = \big\|_{k=1}^{K} \sigma \left(\sum_{j \in \mathcal{N}_i} \alpha_{ij}^k \mathbf{W}^k h_j \right). \tag{4.56}$$

In the final layer of GATs, averaging is used instead of concatenation:

$$\mathbf{h}_i' = \sigma \left(\frac{1}{K} \sum_{k=1}^{K} \sum_{j \in \mathcal{N}_i} \alpha_{ij}^k \mathbf{W}^k \mathbf{h}_j \right). \tag{4.57}$$

4.8.3 Graph Convolutional Networks for VQA

Motivation

VQA tasks often require complicated multimodal reasoning over both objects and their relationships in the image conditioned on the given question. However, only a few methods can effectively model the spatial and semantic relationships between objects. In addition, most VQA models lack interpretability, which is common in deep learning models. To solve these problems, Norcliffe-Brown et al. [31] proposed the graph learner, an interpretable graph convolutional network that can learn complex relationships of objects for VQA tasks.

Method

As shown in Fig. 4.16, given the question embedding $\mathbf{q}$ and detected object features $\mathbf{v}_n$, the graph learner aims at generating an undirected graph $G = \{V, \mathcal{E}, \mathbf{A}\}$, where V is the set of nodes representing the detected objects, $\mathcal{E}$ is the set of edges representing the relationships between detected objects and $\mathbf{A}$ is the corresponding adjacency matrix. Specifically, the graph learner learns the adjacency matrix $\mathbf{A}$ and uses it to construct the set of edges $\mathcal{E}$ and corresponding relationships.

The question embedding $\mathbf{q}$ is first concatenated to the N detected object features $\mathbf{v}_n$. Subsequently, a joint embedding of the question and image features is obtained by a nonlinear function F as

$$\mathbf{e}_n = F([\mathbf{v}_n \| \mathbf{q}]), \qquad n = 1, 2, ..., N, \tag{4.58}$$

where $\|$ represents concatenation. Next, all joint embedding vectors $\mathbf{e}_n$ are concatenated into a joint embedding matrix $\mathbf{E}$, and the adjacency matrix $\mathbf{A}$ is defined as

$$\mathbf{A} = \mathbf{E}\mathbf{E}^T. \tag{4.59}$$

To construct a local connected adjacency matrix in which only the most relevant neighboring nodes are connected by each node, each element in the matrix $\mathbf{E}$ is ranked to be selected or not:

$$\mathcal{N}(i) = topm(\mathbf{a}_i), \tag{4.60}$$

where the *topm* function returns the indices of the elements in the input vector that have the m largest values, and $\mathbf{a}_i$ represents the ith row of the adjacency matrix.

The graph learner utilizes a graph CNN to manage the graph representation of the image. To capture the spatial relationships between two detected objects (nodes i and j), a pairwise pseudocoordinate function centered at i is used, which returns a coordinate vector (ρ, θ) of j, consisting of the orientation θ and distance ρ. For each node i, multiple kernels w_k are used to learn from neighboring nodes $\mathcal{N}(i)$ and generate the graph convolution feature $\mathbf{f}_k(i)$:

$$\mathbf{f}_k(i) = \sum_{j \in \mathcal{N}(i)} w_k(\mathbf{u}(i, j))\mathbf{v}_j \alpha, \qquad k = 1, 2, ..., K, \tag{4.61}$$

where α_{ij} is a scaling weight for each selected element in the adjacency matrix. These convolution features are later concatenated as the final graph representation $\mathbf{H}$, which is passed to the classifier with question embedding q.

$$\mathbf{h}_i = \|_{k=1}^{K} \mathbf{G}_k \mathbf{f}_k(i), \tag{4.62}$$

$$\mathbf{H} = \|_{i=1}^{N} \mathbf{h}_i, \tag{4.63}$$

where $\mathbf{G}_k$ represents the learnable weights.

Performance and Limitation

The graph learner method achieves an overall score of 66.18% on the VQA-v2 test set, which is competitive with the state-of-the-art methods. Qualitative results demonstrate the high interpretability of the proposed model. However, the simple graph structure used in the proposed model cannot effectively manage the more complex relationships between graph items, and a more complex architecture can be utilized to solve this problem. In addition, the performance of the proposed model is dependent on the pretrained object detector, and this aspect can be further enhanced.

4.8.4 Graph Attention Networks for VQA

Motivation

To more accurately answer questions in VQA tasks, the VQA models should capture both the spatial (positional) and semantic (actionable) relationships between objects in the given image rather than merely detecting relevant objects. Thus, Li et al. [20] proposed the relation-aware graph attention network (ReGAT) for VQA tasks, which treats the input images as graphs and captures the complicated relationships between detected objects using graph attention mechanisms.

Method

As shown in Fig. 4.17, four main components are used in ReGAT: an image encoder to generate features of detected objects $\mathcal{V} = \{v_i\}_{i=1}^{K}$ by the Faster R-CNN network, a question encoder to generate question embedding **q** by a bi-directional GRU network with a self-attention mechanism, a relation encoder to model explicit and

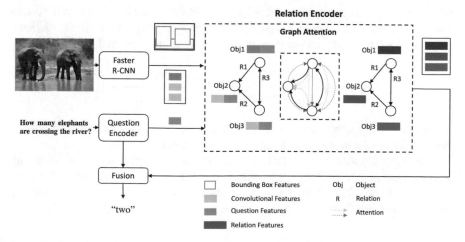

Fig. 4.17 Overview of the relation-aware attention network for VQA

implicit relationships between objects by graph attention mechanisms and a multimodal fusion with the answer predictor.

The first step to build a relation encoder is to construct graphs based on the given image and question. Three graphs are constructed in ReGAT: a fully connected graph $\mathcal{G}_{imp} = \{\mathcal{V}, \mathcal{E}\}$ to model implicit relations and two pruned graphs $\mathcal{G}_{spa}$ and $\mathcal{G}_{sem}$ with prior knowledge to model spatial and semantic relationships, respectively. Among these two pruned graphs, if an explicit relation between two objects does not exist, the edge is pruned. The construction of spatial and semantic graphs can be viewed as a classification task, implemented by pretrained relationship classifiers. Specifically, the spatial relation $spa_{i,j}$ is denoted as $< \texttt{object}_i\texttt{-predicate-object}_j >$, e.g. $< \texttt{kid-cover-sunglasses} >$, and semantic relation $sem_{i,j}$ is denoted as $< \texttt{subject-predicate-object} >$, e.g. $< \texttt{kid-wearing-sunglasses} >$. Note that the relation between objects i and j in spatial and semantic graphs is unsymmetrical.

In the relation encoder, a question-adaptive graph attention mechanism is used. The mechanism is implemented by the concatenation of question embedding $\mathbf{q}$ with K visual features $\mathbf{v}_i$ and a following multihead self-attention mechanism:

$$v_i' = [v_i || q] \quad \text{for } i = 1, \ldots, K, \tag{4.64}$$

$$v_i^\star = \|_{m=1}^M \sigma \left(\sum_{j \in N_i} \alpha_{ij}^m \cdot W^m v_j' \right), \tag{4.65}$$

where M is the number of attention heads, and the attention score α_{ij} varies for implicit and explicit relations. Specifically, for implicit relations, the attention score is defined as

$$\alpha_{ij} = \frac{\alpha_{ij}^b \cdot \exp(\alpha_{ij}^v)}{\sum_{j=1}^K \alpha_{ij}^b \cdot \exp(\alpha_{ij}^v)}, \tag{4.66}$$

where α_{ij}^v and α_{ij}^b represent the similarity between the visual features of objects and relative geometry features of bounding boxes, respectively. For explicit relations, i.e., spatial relations and semantic relations, the attention score is defined as

$$\alpha_{ij} = \frac{\exp((\mathbf{U}v_i')^\top \cdot \mathbf{V}_{dir(i,j)} v_j' + b_{lab(i,j)})}{\sum_{j \in N_i} \exp((\mathbf{U}v_i')^\top \cdot \mathbf{V}_{dir(i,j)} v_j' + b_{lab(i,j)})},$$

where $\mathbf{U}$ and $\mathbf{V}$ represent the projection matrices, and $dir(i, j)$ and $lab(i, j)$ represent the directionality and the label of each edge, respectively.

Finally, $v_i^\star$ is added to v_i as the final relation-aware visual features, which are up with the question embedding q and passed through the answer classifier of a two-layer MLP. Each relation encoder is trained independently, and the predicted answer distributions of each answer classifier are ensembled for inference (Table 4.2).

Table 4.2 Comparison with state-of-the-art methods on VQA v1 dataset

Method	Test-dev				Test-standard			
	Y/N	Num.	Other	All	Y/N	Num.	Other	All
iBOWIMG [50]	76.6	35.0	42.6	55.7	76.8	35.0	42.6	55.9
DPPnet [30]	80.7	37.2	41.7	57.2	80.3	36.9	42.2	57.4
VQA team [4]	80.5	36.8	43.1	57.8	80.6	36.4	43.7	58.2
Neural-Image-QA [28]	78.4	36.4	46.3	58.4	78.2	36.3	46.3	58.4
D-NMN	80.5	43.1	37.4	57.9	–	–	–	58.0
SAN [43]	79.3	36.6	46.1	58.7	–	–	–	58.9
ACK [38]	81.0	38.4	45.2	59.2	81.1	37.1	45.8	59.4
NMN [3]	81.2	38.0	44.0	58.6	81.2	37.7	44.0	58.7
D-NMN [2]	81.1	38.6	45.5	59.4	–	–	–	59.4
DMN+ [39]	80.5	48.3	36.8	60.3	–	–	–	60.4
HieCoAtt [23]	79.7	38.7	51.7	61.8	–	–	–	62.1
MCB-ResNet [10]	82.5	37.6	55.6	64.7	–	–	–	–
MAN [24]	81.5	39.0	54.0	63.8	81.7	37.6	54.7	64.1

Table 4.3 Comparison with state-of-the-art methods on the VQA v2 dataset

Method	Test-dev				Test-standard			
	Y/N	Num.	Other	All	Y/N	Num.	Other	All
MCB [10]	–	–	–	–	78.82	38.28	53.36	62.27
BUTD [1]	81.82	44.21	56.05	65.32	82.20	43.90	56.26	65.67
GCN-VQA [31]	–	–	–	–	82.91	47.13	56.22	66.18
MFH [47]	84.27	50.66	60.50	68.76	–	–	–	66.12
DCN [29]	83.51	46.61	57.26	66.87	–	–	–	66.97
Counter [49]	83.14	51.62	58.97	68.09	–	–	–	68.41
MulRel [5]	84.77	49.84	57.85	68.03	–	–	–	68.41
Pythia [15]	–	–	–	70.01	–	–	–	70.24
BAN [16]	85.42	54.04	60.52	70.04	–	–	–	70.35
DFAF [12]	86.09	53.32	60.49	70.22	–	–	–	70.34
ReGAT [20]	86.08	54.42	60.33	70.27	–	–	–	70.58

Performance and Limitation

The relation-aware graph attention network (ReGAT) method achieves an overall score of 70.58% on the VQA-v2 test set, thereby outperforming the state-of-the-art methods. The ReGAT model is compatible with generic VQA models and can be easily incorporated with state-of-the-art VQA models. With the introduction of ReGAT, several state-of-the-art VQA models exhibit a considerable performance improvement on the VQA-v2 val set. However, the three relations can be used more effectively to solve specific question types (Table 4.3).

References

1. P. Anderson, X. He, C. Buehler, D. Teney, M. Johnson, S. Gould, L. Zhang, Bottom-up and top-down attention for image captioning and visual question answering, in *Proceedings of the IEEE Conference on Computer Vision and Pattern Recognition* (2018), pp. 6077–6086

2. J. Andreas, M. Rohrbach, T. Darrell, D. Klein, Learning to compose neural networks for question answering, in *Proceedings of the Conference of North American Chapter of Association for Computational Linguistics* (2016), pp. 1545–1554

3. J. Andreas, M. Rohrbach, T. Darrell, D. Klein, Neural module networks, in *Proceedings of the IEEE Conference on Computer Vision and Pattern Recognition* (2016), pp. 39–48

4. S. Antol, A. Agrawal, J. Lu, M. Mitchell, D. Batra, C.L. Zitnick, D. Parikh, VQA: visual question answering, in *Proceedings of the IEEE International Conference on Computer Vision* (2015), pp. 2425–2433

5. R. Cadène, H. Ben-younes, M. Cord, N. Thome, MUREL: multimodal relational reasoning for visual question answering, in *Proceedings of the IEEE Conference on Computer Vision and Pattern Recognition* (Computer Vision Foundation/IEEE, 2019), pp. 1989–1998

6. M. Charikar, K.C. Chen, M. Farach-Colton, Finding frequent items in data streams, in *Automata, Languages and Programming*, Lecture Notes in Computer Science, vol. 2380 (Springer, 2002), pp. 693–703

7. K. Chen, J. Wang, L. Chen, H. Gao, W. Xu, R. Nevatia, ABC-CNN: an attention based convolutional neural network for visual question answering. CoRR, arXiv:1511.05960 (2015)

8. M. de Marneffe, C.D. Manning, The Stanford typed dependencies representation (2008), pp. 1–8

9. J. Dong, X. Li, C.G.M. Snoek, Predicting visual features from text for image and video caption retrieval. IEEE Trans. Multimedia **20**(12), 3377–3388 (2018)

10. A. Fukui, D.H. Park, D. Yang, A. Rohrbach, T. Darrell, M. Rohrbach, Multimodal compact bilinear pooling for visual question answering and visual grounding, in *Proceedings of the Conference on Empirical Methods in Natural Language Processing*, arXiv:1606.01847 (2016), pp. 457–468

11. H. Gao, J. Mao, J. Zhou, Z. Huang, L. Wang, W. Xu, Are you talking to a machine? dataset and methods for multilingual image question, in *Proceedings of the Advances in Neural Information Processing Systems* (2015), pp. 2296–2304

12. P. Gao, Z. Jiang, H. You, P. Lu, S.C. H. Hoi, X. Wang, H. Li, Dynamic fusion with intra- and inter-modality attention flow for visual question answering, in *Proceedings of the IEEE Conference on Computer Vision and Pattern Recognition* (Computer Vision Foundation/IEEE, 2019), pp. 6639–6648

13. Y. Goyal, T. Khot, D. Summers-Stay, D. Batra, D. Parikh, Making the V in VQA matter: elevating the role of image understanding in visual question answering, in *Proceedings of the IEEE Conference on Computer Vision and Pattern Recognition* (IEEE Computer Society, 2017), pp. 6325–6334

14. A. Jiang, F. Wang, F. Porikli, Y. Li, Compositional memory for visual question answering. CoRR, arXiv:1511.05676 (2015)

15. Y. Jiang, V. Natarajan, X. Chen, M. Rohrbach, D. Batra, D. Parikh, Pythia v0.1: the winning entry to the VQA challenge 2018. CoRR, arXiv:1807.09956 (2018)

16. J. Kim, J. Jun, B. Zhang, Bilinear attention networks, in *Proceedings of the Advances in Neural Information Processing Systems*, ed. by S. Bengio, H.M. Wallach, H. Larochelle, K. Grauman, N. Cesa-Bianchi, R. Garnett (2018), pp. 1571–1581

17. J. Kim, K. W. On, W. Lim, J. Kim, J. Ha, B. Zhang, Hadamard product for low-rank bilinear pooling, in *Proceedings of the International Conference on Learning Representations*. OpenReview.net (2017)

18. T.N. Kipf, M. Welling, Semi-supervised classification with graph convolutional networks, in *Proceedings of the International Conference on Learning Representations* (2017)

19. R. Krishna, Y. Zhu, O. Groth, J. Johnson, K. Hata, J. Kravitz, S. Chen, Y. Kalantidis, L. Li, D.A. Shamma, M.S. Bernstein, L. Fei-Fei, Visual genome: connecting language and vision using crowdsourced dense image annotations. Int. J. Comput. Vis. **123**(1), 32–73 (2017)
20. L. Li, Z. Gan, Y. Cheng, J. Liu, Relation-aware graph attention network for visual question answering, in *Proceedings of the IEEE International Conference on Computer Vision* (2019), pp. 10312–10321
21. X. Li, S. Jiang, Know more say less: image captioning based on scene graphs. IEEE Trans. Multimedia **21**(8), 2117–2130 (2019)
22. T. Lin, M. Maire, S.J. Belongie, J. Hays, P. Perona, D. Ramanan, P. Dollár, C.L. Zitnick, Microsoft COCO: common objects in context, in *Proceedings of the European Conference on Computer Vision*, vol. 8693 (Springer, 2014), pp. 740–755
23. J. Lu, J. Yang, D. Batra, D. Parikh, Hierarchical question-image co-attention for visual question answering, in *Proceedings of the Advances in Neural Information Processing Systems* (2016), pp. 289–297
24. C. Ma, C. Shen, A.R. Dick, Q. Wu, P. Wang, A. van den Hengel, I.D. Reid, Visual question answering with memory-augmented networks, in *Proceedings of the IEEE Conference on Computer Vision and Pattern Recognition* (2018), pp. 6975–6984
25. L. Ma, Z. Lu, H. Li, Learning to answer questions from image using convolutional neural network, in *Proceedings of the Conference on AAAI* (2016), pp. 3567–3573
26. M. Malinowski, M. Fritz, A multi-world approach to question answering about real-world scenes based on uncertain input, in *Proceedings of the Advances in Neural Information Processing Systems* (2014), pp. 1682–1690
27. M. Malinowski, M. Rohrbach, M. Fritz, Ask your neurons: a neural-based approach to answering questions about images, in *Proceedings of the IEEE International Conference on Computer Vision* (2015), pp. 1–9
28. M. Malinowski, M. Rohrbach, M. Fritz, Ask your neurons: a neural-based approach to answering questions about images, in *Proceedings of the IEEE International Conference on Computer Vision* (2015), pp. 1–9
29. D. Nguyen, T. Okatani, Improved fusion of visual and language representations by dense symmetric co-attention for visual question answering, in *Proceedings of the IEEE Conference on Computer Vision and Pattern Recognition* (Computer Vision Foundation/IEEE Computer Society, 2018), pp. 6087–6096
30. H. Noh, P.H. Seo, B. Han, Image question answering using convolutional neural network with dynamic parameter prediction, in *Proceedings of the IEEE Conference on Computer Vision and Pattern Recognition* (IEEE Computer Society, 2016), pp. 30–38
31. W. Norcliffe-Brown, S. Vafeias, S. Parisot, Learning conditioned graph structures for interpretable visual question answering, in *Proceedings of the Advances in Neural Information Processing Systems* (2018)
32. M. Ren, R. Kiros, R.S. Zemel, Exploring models and data for image question answering, in *Proceedings of the Advances in Neural Information Processing Systems* (2015), pp. 2953–2961
33. A. Singh, V. Natarajan, M. Shah, Y. Jiang, X. Chen, D. Batra, D. Parikh, M. Rohrbach, Towards vqa models that can read, in *Proceedings of the IEEE Conference on Computer Vision and Pattern Recognition* (2019), pp. 8317–8326
34. A. Vaswani, N. Shazeer, N. Parmar, J. Uszkoreit, L. Jones, A.N. Gomez, L. Kaiser, I. Polosukhin, Attention is all you need, in *Proceedings of the Advances in Neural Information Processing Systems* (2017), pp. 5998–6008
35. P. Velickovic, G. Cucurull, A. Casanova, A. Romero, P. Liò, Y. Bengio, Graph attention networks, in *Proceedings of the International Conference on Learning Representations* (2018)
36. B. Wang, Y. Xu, Y. Han, R. Hong, Movie question answering: remembering the textual cues for layered visual contents, in *Proceedings of the Conference on AAAI* (2018), pp. 7380–7387
37. P. Wang, Q. Wu, C. Shen, A.R. Dick, A. van den Hengel, Explicit knowledge-based reasoning for visual question answering, in *Proceedings of the International Joint Conference on Artificial Intelligence*, ed. by C. Sierra (2017), pp. 1290–1296

38. Q. Wu, P. Wang, C. Shen, A.R. Dick, A. van den Hengel, Ask me anything: free-form visual question answering based on knowledge from external sources, in *Proceedings of the IEEE Conference on Computer Vision and Pattern Recognition* (2016), pp. 4622–4630
39. C. Xiong, S. Merity, R. Socher, Dynamic memory networks for visual and textual question answering, in *Proceedings of the International Conference on Machine Learning* (2016), pp. 2397–2406
40. H. Xu, K. Saenko, Ask, attend and answer: exploring question-guided spatial attention for visual question answering, in *Proceedings of the European Conference on Computer Vision*, vol. 9911 (2016), pp. 451–466
41. K. Xu, J. Ba, R. Kiros, K. Cho, A.C. Courville, R. Salakhutdinov, R.S. Zemel, Y. Bengio, Show, attend and tell: neural image caption generation with visual attention, in *Proceedings of the International Conference on Machine Learning*, vol. 37 (2015), pp. 2048–2057
42. N. Xu, H. Zhang, A. Liu, W. Nie, Y. Su, J. Nie, Y. Zhang, Multi-level policy and reward-based deep reinforcement learning framework for image captioning. IEEE Trans. Multimedia **22**(5), 1372–1383 (2020)
43. Z. Yang, X. He, J. Gao, L. Deng, A.J. Smola, Stacked attention networks for image question answering, in *Proceedings of the IEEE Conference on Computer Vision and Pattern Recognition* (2016), pp. 21–29
44. T. Yao, Y. Pan, Y. Li, T. Mei, Hierarchy parsing for image captioning. Int. J. Comput. Vis. 2621–2629 (2019)
45. L. Yu, E. Park, A.C. Berg, T.L. Berg, Visual madlibs: fill in the blank description generation and question answering, in *Proceedings of the IEEE International Conference on Computer Vision* (2015), pp. 2461–2469
46. Z. Yu, D. Xu, J. Yu, T. Yu, Z. Zhao, Y. Zhuang, D. Tao, Activitynet-qa: a dataset for understanding complex web videos via question answering, in *Proceedings of the Conference on AAAI* (2019), pp. 9127–9134
47. Z. Yu, J. Yu, C. Xiang, J. Fan, D. Tao, Beyond bilinear: generalized multimodal factorized high-order pooling for visual question answering. IEEE Trans. Neural Netw. Learn. Syst. **29**(12), 5947–5959 (2018)
48. P. Zhang, Y. Goyal, D. Summers-Stay, D. Batra, D. Parikh, Yin and yang: balancing and answering binary visual questions, in *Proceedings of the IEEE Conference on Computer Vision and Pattern Recognition* (2016), pp. 5014–5022
49. Y. Zhang, J.S. Hare, A. Prügel-Bennett, Learning to count objects in natural images for visual question answering, in *Proceedings of the International Conference on Learning Representations*
50. B. Zhou, Y. Tian, S. Sukhbaatar, A. Szlam, R. Fergus, Simple baseline for visual question answering. CoRR, arXiv:1512.02167 (2015)
51. Y. Zhu, O. Groth, M.S. Bernstein, L. Fei-Fei, Visual7w: grounded question answering in images, in *Proceedings of the IEEE Conference on Computer Vision and Pattern Recognition* (2016), pp. 4995–5004

Chapter 5
Knowledge-Based VQA

Abstract Tasks such as VQA often require common sense and factual information in addition to the information learned from a task-specific dataset. Therefore, a knowledge-based VQA task is established. In this chapter, we first introduce the main datasets proposed for knowledge-based VQA and knowledge bases such as DBpedia and ConceptNet. Subsequently, we classify methods from three aspects: knowledge embedding, question-to-query translation and querying knowledge base methods.

5.1 Introduction

The VQA task aims to understand the content of an image and answer questions, often requiring prior nonvisual information. In real life, humans tend to combine visual observation with external knowledge when answering questions. Therefore, the model must refer to information that the image itself does not contain, such as external or common-sense knowledge. However, the existing VQA models cannot derive additional knowledge from existing datasets. Therefore, knowledge-based visual question answering has been proposed. Knowledge-based VQA requires external knowledge beyond the visual content to answer questions regarding images, which is challenging but essential for the implementation of universal visual question answering. Since the structured representation of knowledge has been extensively studied, external knowledge can be referred to as the knowledge base. Many researchers [14, 23–25] have focused on knowledge-based VQA. In this chapter, we examine knowledge-based VQA from three aspects: datasets, knowledge bases and methods. The dataset introduces four kinds of mainstream datasets: KB-VQA, FVQA, OK-VQA and KRVQA. Subsequently, we review the methods associated with knowledge-based VQA, which can be categorized as methods for knowledge embedding (Sect. 5.4), question-to-query translation (Sect. 5.5) and querying of knowledge bases (Sect. 5.6).

5.2 Datasets

Many datasets have been proposed for research on knowledge-based VQA. In the following sections, we describe the existing knowledge-based VQA datasets, specify the methods to create such datasets and compare the datasets. The key characteristics are summarized in Table 5.1.

KB-VQA

The KB-VQA dataset [23] aims to evaluate the ability of the VQA model to answer questions pertaining to a high knowledge level and reason using external knowledge.

This dataset involves 700 images from the MSCOCO dataset validation set and 3 to 5 question and answer pairs for each image, resulting in 2,402 questions. Each question in the dataset has been generated by humans based on 23 predefined templates. For example, the template for *IsThereAny* is "Is there any <concept>?".

Compared to those in the other VQA datasets, questions in the KB-VQA dataset generally require a higher level of external knowledge to answer. The questions are associated with three labels: "visual", "common-sense" and "KB-knowledge". "Visual" questions are answered directly through visual concepts of ImageNet and MSCOCO ("Is there a car in the image?"), "Common-sense" questions do not require adults to refer to external knowledge ("How many dogs are in the image?") and "KB-knowledge" questions are answered using knowledge bases, such as Wikipedia ("What do the animals in the picture and zebras have in common?").

FVQA

The FVQA dataset [24] provides a supporting fact for question and answer pairs in the form of a structural triplet image-question-answer-supporting fact. For example, for the question "Which animal in the image is able to climb trees?", the answer is "cat", and the supporting facts are <Cat, CapableOf, ClimbingTrees>.

FVQA consists of 2,190 images, 5,286 questions and a knowledge base of 193,449 facts. FVQA builds the knowledge base by collecting knowledge triples of knowledge bases: DBpedia [2], WebChild [21, 22] and ConceptNet [19]. The dataset consists of

Table 5.1 Characteristics of major datasets for knowledge-based VQA

Dataset	Number of images	Number of questions	Number question of categories	Avg. question length	Avg. ans. length
KB-VQA [23]	700	2,402	23	6.8	2.0
FVQA [24]	2,190	5,826	32	9.5	1.2
OK-VQA [13]	14,031	14,055	10	8.1	1.3
KRVQA [5]	32,910	157,201	–	11.7	–

5 train-test folds. FVQA has 32 types of questions, which are categorized according to the type of visual concept (object, scene, or action), source of the answer (image or knowledge base) and knowledge base supporting the facts (DBpedia, WebChild, or ConceptNet).

When creating the FVQA dataset, the annotator selects the image and visual elements of the image and subsequently selects a pre-extracted supporting fact related to the visual concept. Finally, the annotator specifies a question/answer related to the selected supporting facts.

By providing supporting facts, FVQA enables complex questions to be answered, even if all the required information is not shown in the image. Moreover, the dataset supports explicit reasoning in a question and answer. Specifically, this framework indicates how a method might arrive at an answer. This information can be used for answer reasoning, searching for other appropriate facts, or evaluating answers that contain an inference chain.

OK-VQA

The outside knowledge VQA (OK-VQA) dataset [13] consists of 14,031 images and 14,055 questions and 7,178 unique question words, covering a variety of knowledge categories, including science and technology, history and sports. OK-VQA uses random images from the MSCOCO dataset, using the original 80k training and 40k validation split to split the training and testing datasets. Unlike the existing fact-based VQA datasets, such as KB-VQA and FVQA, which require the VQA system to achieve visual reasoning with a given knowledge base, OK-VQA requires reasoning based on uninstructed knowledge. In addition, each question is labeled considering one of 10 knowledge categories: vehicles and transportation (VT); brands, companies and products (BCP); objects, materials and clothing (OMC); sports and recreation (SR); cooking and food (CF); geography, history, language and culture (GHLC); people and everyday life (PEL), plants and animals (PA); science and technology (ST); and weather and climate (WC). If a question does not fit into any category, then it is classified as "Other".

KRVQA

The knowledge-routed visual question reasoning (KRVQA) dataset [5] is the first large-scale dataset that requires knowledge reasoning on natural images. The dataset consists of 32,910 images, 157,201 pairs of different types of questions and answers and 194,449 knowledge triplets. The average length of the questions is 11.7 words. Questions can be divided into one- and two-step questions according to the reasoning steps and KB-related and KB-unrelated according to the involved knowledge.

The construction of the KRVQA dataset is based on the scene graph annotations of the visual genome dataset [10] and knowledge base of the FVQA dataset. To generate nonbiased answer pairs, the KRVQA first cleans up object and relation names in the visual genome scene map annotations. Using the scene graph and related knowledge triplets, an image-specific knowledge graph is formed, which is used to describe the objects, relations and knowledge related to the images. Subsequently, facts are extracted from the graph and assembled into a reasoning program. Finally, the question and answer pairs are generated based on the program layout and predefined question template.

5.3   Knowledge Bases

5.3.1   DBpedia

The DBpedia project is a data corpus from Wikipedia that aims at extracting structured information from Wikipedia and making this information available on the web. DBpedia allows users to ask complex queries on datasets from Wikipedia and link other datasets on the web to Wikipedia data. The project extracts knowledge from Wikipedia in 111 different languages. The project was started in 2007 by Soren Auer and Jens Lehmann [2].

Wikipedia is the most widely used encyclopedia, with official editions in 287 languages, ranging in size from a few hundred articles to 3.8 million articles (in English). However, similar to many other web applications, the search functionality of Wikipedia is limited to full-text search, which leads to its extremely limited value. Moreover, the platform has certain other drawbacks, such as conflicting data, inconsistent classification conventions, errors and spam.

Thus, the DBpedia project focuses on transforming Wikipedia content into structured knowledge so that semantic web technologies can perform complex queries against Wikipedia, link it to other datasets on the web, or create new applications or mashups.

The largest DBpedia knowledge base has been extracted from the English version of Wikipedia. The dataset contains information on more than 400 million facts about 3.7 million things. The DBpedia knowledge base has been extracted from 110 Wikipedia versions, containing 1.46 billion facts about 10 million things. The DBpedia project maps Wikipedia information boxes from 27 language versions into a shared ontology consisting of 320 classes and 1,650 attributes.

Notably, DBpedia has been created by extracting structured information from Wikipedia and is thus significantly larger and more general than a handcrafted KB. The whole DBpedia dataset describes 4.58 million entities, among which 4.22 million entities are classified in a consistent ontology. DBpedia concepts are described with short and long abstracts in 13 languages.

DBpedia is available on the web in three forms. First, the base is available as a downloadable dataset. Second, DBpedia provides services through public SPARQL endpoints, and third, it provides dereferenceable URIs based on the linked data principle.

5.3.2 ConceptNet

ConceptNet is a knowledge graph that connects words and phrases of natural language with labeled, weighted edges. The corpus contains over 21 million edges and 8 million nodes. The corresponding English vocabulary contains approximately 1.5 million nodes, with at least 10,000 nodes in 83 languages.

The initial version of ConceptNet was proposed in 2004 [12], and ConceptNet 5.5 [19] was proposed in 2015. The dataset is built from diverse sources, such as Open Mind Common Sense (OMCS), Wiktionary, "games with a purpose", Open Multilingual WordNet, JMDict (Breen 2004), OpenCyc and a subset of DBPedia. ConceptNet's largest source of input is Wiktionary, which offers 18.1 million edges. Most of the characters of ConceptNet pertain to OMCS and the various games with a purpose. Compared to other knowledge base resources, ConceptNet provides an adequately large and free knowledge graph that focuses on the common-sense meaning of words.

Notably, ConceptNet represents relationships between words, which can be simply expressed as a triplet of their start node, relation label (such as IsA and UsedFor), and end node. For example, the assertion "a cat has a tail" can be expressed as (cat, HasA, tail).

5.4 Knowledge Embedding Methods

5.4.1 Word-to-Vector Representation

Motivation. In the traditional approaches for knowledge-based VQA, the model first extracts visual features from the given image and linguistic features from the question, and these features are associated with the external knowledge base. To search for relevant facts from the knowledge base, the model predicts attributes from images or relation types from questions. After the external knowledge is retrieved, it must be encoded. To this end, the most commonly applied approach is word-to-vector representation, such as GloVe embedding or Doc2Vec. Finally, the information is summarized to obtain the final answer (see Fig. 5.1).

Methods. Wu et al. [25] proposed a method to combine the representation of image content with information extracted from a common knowledge base to answer a wide range of image-based questions. In the current model, the author first uses a CNN to produce the attribute-based representation from the image. Based on the image attributes, the model generates image captions as internal representations. Subsequently, the model uses SPARQL to retrieve relevant knowledge from external knowledge according to the predicted attributes. Since the text returned by the SPARQL query is usually considerably longer than the generated captions, the model uses Doc2vec to extract semantic meaning from the retrieved knowledge paragraphs. Specifically, Doc2vec, also referred to as paragraph vectors, is an unsupervised

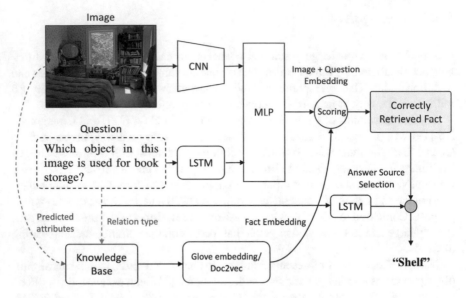

Fig. 5.1 Overview of the word-to-vector representation method

algorithm that learns fixed-length feature representations from variable-length fragments of text, such as sentences, paragraphs and documents. Finally, the predicted
attribute, captions and database-based knowledge vector are passed as the input to
an LSTM that learns to predict the answer to the input question in the form of a
sequence of words.

Narasimhan et al. [14] developed a learning-based retrieval method that directly
learns an embedding of facts and question-and-image pairs into a space. This model
avoids the generation of an explicit query and learns to transform extracted visual
concepts into a vector that is close to the relevant facts in the embedded space
of learning. Specifically, the model extracts image features from a CNN and text
features from an LSTM. Subsequently, the model uses a multilayer perceptron (MLP)
to combine the two modalities. An LSTM is used to predict the fact relation type
from the question and retrieve facts from the fact knowledge base. The retrieved
structured facts are encoded using GloVe-100 embedding. Finally, the retrieved facts
are ranked with image + question + visual concept embedding, and the top-ranked
facts are returned.

Performance and Limitations. Tables 5.2, 5.3 and 5.4 summarize the performances
of all the discussed methods and datasets. Wu et al. evaluated the Toronto COCO-QA
and VQA datasets and demonstrated that the use of external knowledge base lessons
can effectively enhance the performance. However, the method proposed by Wu et
al. did not perform any explicit reasoning and may have ignored possible structures
in the knowledge base. In addition, this method only extracts discrete text fragments
from the knowledge base, thereby ignoring the structural representation ability. The

Table 5.2 Comparison of results on the FVQA dataset

Model	Overall top-1	Accuracy (%) top-3
LSTM-Question+Image+Pre-VQA	24.98	40.40
Hie-Question+Image+Pre-VQA	43.14	59.44
FVQA (top-3-QQmaping) [24]	56.91	64.65
FVQA (Ensemble) [24]	58.76	–
Straight to the Facts (STTF) [14]	62.20	75.60
Reading comprehension	62.96	70.08
Out of the box (OB) [15]	69.35	80.25
Mucko [29]	73.06	85.94
GRUC [26]	79.63	91.20

approach proposed by Narasimhan et al. was evaluated on the FVQA dataset, and the performance of this method was considered highest in the top-1 accuracy metric.

5.4.2   Bert-Based Representation

Motivation. Pretrained language representation models such as BERT are undergoing rapid advancement. However, in the knowledge-based VQA task, most of the existing studies are based on context-free word embedding rather than the fusion of a knowledge graph (KG) and image representation.

Methods. Garderes et al. proposed ConceptBert [7], which uses pretrained image and language features and fuses them with KG embeddings to capture image-and-question knowledge-specific interactions. As shown in Fig. 5.2, ConceptBert consists of visual embedding, text embedding and knowledge graph representation. Visual representation is obtained using the Faster R-CNN framework [17]. Question representation is realized using BERT [6]. ConceptNet is used as the knowledge base. The method uses a graph convolutional network to integrate information from the local neighborhood of the node in the graph. The network consists of an encoder and a decoder. Graph convolution encoders take graphs as the input and encode each node. The encoder operates by sending messages from a node to its neighbors and weighting them according to the relationship types defined by the edges. This operation occurs in multiple layers and contains multiple hops of information from a node. The representation of the last layer is embedded as a graph of nodes. The vision-language module represents the joint embeddings of the language and visual content, which is based on two parallel BERT-style streams. The concept-language module represents the language features conditioned on knowledge graph embeddings, which is a series of transformer blocks that examine question tokens based on

Table 5.3 Comparison of results on the KRVQA dataset

Method	KB-unrelated							KB-related					Overall
	One-step			Two-step				One-step		Two-step			
	0	1	2	3	4	5	6	2	3	4	5	6	
Q-type	36.19	2.78	8.21	33.18	35.97	3.66	8.06	0.09	0.00	0.18	0.06	0.33	8.12
LSTM	45.98	2.79	2.75	43.26	40.67	2.62	1.72	0.43	0.00	0.52	1.65	0.74	8.81
Program predict	58.86	50.98	59.17	54.71	57.31	54.17	57.64	65.16	33.95	71.64	63.05	76.53	61.62
FiLM [16]	52.42	21.35	18.50	45.23	42.36	21.32	15.44	6.27	5.48	4.37	4.41	7.19	16.89
MFH [27]	43.74	28.28	27.49	38.71	36.48	20.77	21.01	12.97	5.10	6.05	5.02	14.38	19.55
UpDn [1]	56.42	29.89	28.63	49.69	43.87	24.71	21.28	11.07	8.16	7.09	5.37	13.97	21.85
MCAN [28]	49.60	27.67	25.76	39.69	37.92	21.22	18.63	12.28	9.35	9.22	5.23	13.34	20.52

Table 5.4 Comparison of results on the OK-VQA dataset

Model	Overall top-1	Accuracy (%) top-3	VT	BCP	OMC	SR	CF	GHLC	PEL	PA	ST	WC	Other
Q-Only [13]	14.93	–	14.64	14.19	11.78	15.94	16.92	11.91	14.02	14.28	19.76	25.74	13.51
MLP [13]	20.67	–	21.33	15.81	17.76	24.69	21.81	11.91	17.15	21.33	19.29	29.92	19.81
BAN [9]	25.17	–	23.79	17.67	22.43	30.58	27.90	25.96	20.33	25.60	20.95	40.16	22.46
MUTAN [3]	26.41	–	25.36	18.95	24.02	33.23	27.73	17.59	20.09	30.44	20.48	39.38	22.46
ArticleNet (AN) [13]	5.28	–	4.48	0.93	5.09	5.11	5.69	5.24	3.13	6.95	5.00	9.92	5.33
BAN+AN [13]	25.61	–	24.45	19.88	21.59	30.79	29.12	20.57	21.54	26.42	27.14	38.29	22.16
MUTAN+AN [13]	27.84	–	25.56	23.95	26.87	33.44	29.94	20.71	25.05	29.70	24.76	39.84	23.62
BAN/AN oracle [13]	27.59	–	26.35	18.26	24.35	33.12	30.46	28.51	21.54	28.79	24.52	41.4	25.07
Mucko [29]	29.20	30.66	–	–	–	–	–	–	–	–	–	–	–
GRUC [26]	29.87	32.65	29.84	25.23	30.61	30.92	28.01	26.24	29.21	31.27	27.85	38.01	26.21
ConceptBert [7]	33.66	–	30.38	28.02	30.65	37.85	35.08	32.91	28.55	35.88	32.38	47.13	31.47

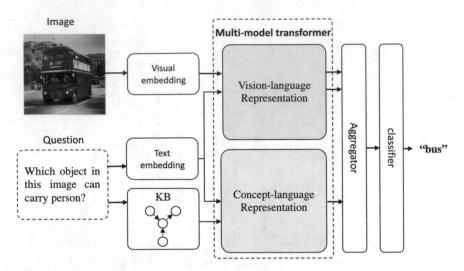

Fig. 5.2 Overview of the BERT-based representation method

the KG embeddings. Finally, the concept-vision-language module uses the compact trilinear interaction (CTI) to generate a joint representation. In addition, ConceptBert does not need external knowledge annotations or search queries.

Recently, Shevchenko et al. [18] proposed a general-purpose technique that injects additional information from the knowledge base into the visual and language transformer. The method preprocesses the knowledge base into knowledge embedding. Moreover, the method uses an auxiliary objective to align the representation of its learning with the knowledge embedding. This approach is implemented in addition to LXMERT [20], which is a state-of-the-art multitasking model.

Performance and limitations. ConceptBert has been evaluated on the VQA 2.0 dataset [8] and OK-VQA dataset. Concepts related to "cooking and food" (CF), "plants and animals" (PA) and "science and technology" (ST) correspond to a superior performance in the OK-VQA dataset. This phenomenon likely occurs because the answers to such questions are usually entities different than the main entity in the question and visual features in the image. This aspect indicates that the information extracted from the knowledge graph is of significance in determining the answer. An extensive empirical evaluation of Shevchenko's approach on four downstream tasks demonstrates that the method performs satisfactorily in knowledge-based VQA (OK-VQA and FVQA datasets) and visual reasoning (NLVR2 and SNLI-VE datasets) tasks.

5.5 Question-to-Query Translation

When querying external knowledge, the model must select an entity from the knowledge base to perform auxiliary reasoning to obtain the final answer. Generally, there exist two kinds of methods to realize question-to-query translation: query-mapping methods and learning-based methods.

5.5.1 Query-Mapping-Based Methods

Motivation. To translate questions into queries, query-mapping-based methods normally parse questions into keywords and retrieve them from the supporting entity.

Ahab [23] detects relevant content from images and searches information from a knowledge base. To obtain the query, Ahab first parses the question into keywords, retrieves relevant facts through keyword matching and finally predicts the answer. Ahab reduces the question to one of the available query templates, which parse questions using NLP tools. Specifically, the natural language toolkit (NLTK) is used to tag each word in the question, which consists of a tokenizer, a part-of-speech tagger and a lemmatizer. Subsequently, the tagged question is parsed by a set of regular expressions (regex), where each regex is defined for a specific question template. The extracted slot phrases are mapped to KB entities, and the appropriate SPARQL queries are formed based on the question template.

FVQA [23] automatically classifies and maps the question to a query that is not subject to the constraint of template. As shown in Fig. 5.3, the KB query is implemented based on three attributes of the question: visual concepts, predicates and answer sources. Overall, there are 32 combinations of the three properties. Each combination is considered a query type, and the LSTM model is used to learn a 32-class classifier to identify the three attributes of the input question and execute a specific query.

Cao et al. [4] proposed a knowledge-routed modular network (KM-net), which performs multistep reasoning by incorporating visual knowledge and common-sense knowledge. For a given question, KM-net parses the question into the query layout via a query estimator. The query estimator adopts the widely used sequence-to-sequence model, which takes the sequence of words in the question as the input and predicts the sequence of query tokens.

Performance and limitations. Ahab [23] relies heavily on predefined templates and accepts only predefined format problems. Although FVQA reduces the question to a query template, the types of questions that can be asked are limited, especially when the question does not accurately refer to a visual concept or information included in a knowledge base. The accuracy and interpretation capability of KM-net have been evaluated on the HVQR dataset.

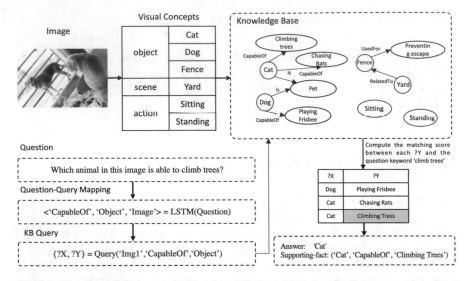

Fig. 5.3 Overview of the query-mapping-based method. First, the visual concepts of the input images are extracted through and linked to the corresponding semantic entities in the knowledge base. The input question is first mapped to the query type by using the LSTM model, from which the type of predicate, visual concept and answer source can be determined. Subsequently, a specific query is executed to identify all facts meeting the search conditions in KB. These facts further match the keywords extracted from the question sentences. Subsequently, the model selects the fact that matches the highest score to obtain the corresponding answer

5.5.2 Learning-Based Methods

Motivation. The drawback of query mapping is that it does not focus on the most notable visual concepts and exhibits an inferior performance in the presence of synonyms and homographs. Therefore, a learning-based approach is proposed that can embed image question pairs and facts in the same space and sort them according to the relevance of the facts.

Methods. The method proposed by Narasimhan et al. [15] introduces visual information into the fact graph and uses implicit graph reasoning to predict answers. In particular, the method applies GCN on the fact graph, and each node is represented by a fixed image problem entity embedded form. This method provides the visual information equally to each graph node through the connection of the image, problem and entity embeddings. Only part of the visual content is relevant to the problem and entity. In addition, because each node is represented in a fixed image problem entity embedding form, the fact graph remains isomorphic, which limits the flexibility of the model to adaptively capture evidence from different patterns.

Zhu et al. [29] proposed a model named Mucko that focuses on multilayer cross-modal knowledge reasoning. Mucko consists of two modules: multimodal heterogeneous graph construction and cross-modal heterogeneous graph reasoning. Multimodal heterogeneous graph construction encodes images through three layers of graphs: visual graph, semantic graph and fact graph. The visual layer preserves the appearance of objects and their relationships, the semantic layer provides a high-level abstraction for connecting visual and factual information and the fact layer supports the corresponding factual knowledge. Furthermore, the authors proposed a modality-aware heterogeneous graph convolutional network to capture problem-oriented evidence from different modalities. This network contains two parts: intramodal knowledge selection to choose question-oriented knowledge from each layer of graphs by intramodal graph convolutions and cross-modal knowledge reasoning to realize the adaptive selection of complementary evidence in three-level graphs through cross-modal graph convolutions.

Yu et al. [26] interpreted images using a multimodal knowledge graph and adopted a memory-based recurrent network for cross-modal reasoning to obtain complementary evidence from different modalities. The proposed model consists of four modules: multimodal heterogeneous graph construction, intramodal knowledge selection, cross-modal knowledge reasoning and global assessment and answer prediction. Specifically, the multimodal knowledge graph construction module represents knowledge of different modalities through different knowledge graphs, including the visual, semantic and fact graphs. The intramodal knowledge selection module selects the knowledge related to the problem from each mode of the knowledge graph. The cross-modal knowledge reasoning module iteratively collects complementary evidence from visual and semantic knowledge graphs via the graph-based read, update and control (GRUC) module. The global assessment and answer prediction module uses graph convolutional networks to jointly analyze all concepts and a binary classifier to predict the answer.

Performance and limitations. Narasimhan's approach exhibits excellent performance on FVQA datasets and does not require visual concept types or answer sources. These improvements can be attributed to the joint reasoning pertaining to answers, which helps share information before a final decision is made. However, since the visual information is fully provided, redundant information may be introduced to infer the answer. Moreover, each node is represented in a fixed image-question-entity embedding form, and the fact graph is isomorphic, which limits the flexibility of the model to adaptively capture evidence from different modalities.

GRUC achieves state-of-the-art performance on three benchmark datasets—FVQA, Visual7W-KB and OK-VQA—and its effectiveness and interpretability have been proven. This model exhibits satisfactory explanatory abilities and can identify the knowledge selection patterns under different modes through comprehensive visualization.

5.6   Methods to Query Knowledge Bases

5.6.1   RDF Query

Motivation. The information in KBs can be efficiently accessed using a query language. In structured KBs, knowledge is typically represented by a large number of triples of the form (arg1, rel, arg2), where arg1 and arg2 denote two concepts in the KB, and rel is a predicate representing the relationship between the concepts. A collection of such triples forms a large interlinked graph. Such triples are often described according to a resource description framework (RDF) specification and housed in a relational database management system (RDBMS) or triple store.

The resource description framework (RDF) is the standard format for the knowledge base, having the form $f_i = (a_i, r_i, b_i)$, where a_i is a visual concept in the image, b_i is an attribute or phrase and r_i is a relation between the two entities. For example, the information that "The image contains a cat object" can be expressed as (Img, contain, Obj-1) and (Obj-1, name, ObjCat-cat).

Methods. Ahab [23] detected visual concepts from images and stored them as RDF triples. As shown in Fig. 5.4, by mapping object/property/scene categories to

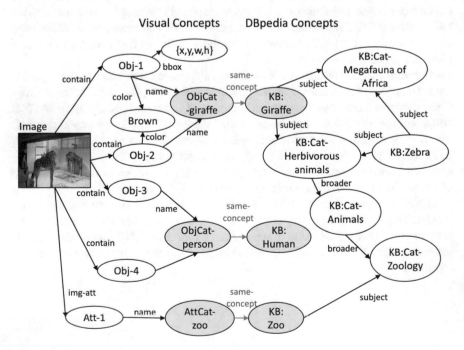

Fig. 5.4 RDF-query visualization. Using the predicate *same concept* to map the object, attribute and scene categories to DBpedia entities, the extracted visual concept graph (left) is linked to DBpedia (right)

DBpedia entities, these visual concepts are linked to the external knowledge base that has the same semantic meaning. Thus, the resulting RDF graph includes all relevant information in DBpedia that corresponds to the visual concepts. Finally, the combination of this image and DBpedia information is accessed through a local Openlink Virtuoso RDBMS.

Wu et al. [25] adopted an SQL-like query language for RDF, SPARQL, to access the knowledge base. Given an image and its prediction attributes, the method uses the top five strongest predicted attributes to generate a DBpedia query. The comment field text for each query is retrieved.

5.6.2 Memory Network Query

Motivation. The existing approaches typically use structured knowledge graphs and images based on supporting facts for reasoning. These algorithms first extract visual concepts from a given image and explicitly implement inference on a structured knowledge base. However, it is not easy to extract adequate visual information due to the lack of structure as well as the grammatical structure, like language conventions.

Methods. Li et al. [11] proposed a knowledge-incorporated dynamic memory network framework (KDMN), which uses dynamic memory networks to introduce a large amount of external knowledge to answer the visual questions of open domains. KDMN is the first attempt at combining external knowledge and image representation with memory mechanisms. As shown in Fig. 5.5, KDMN is composed of three parts: candidate knowledge retrieval, dynamic memory network and knowledge-incorporated open-domain VQA. First, the candidate knowledge retrieval module retrieves candidate knowledge related to the images and questions. To extract the candidate nodes from ConceptNet, Fast R-CNN is used to extract visual objects from images and syntax analysis is performed to extract textual keywords from questions. The candidate knowledge is represented as context-relevant knowledge triples. Subsequently, the image representation and knowledge are extracted and integrated into a common space and stored in the dynamic storage module. Unlike ordinary RDF queries, KDMN generates the query vector by feeding visual and textual features into a nonlinear fully connected layer to capture information from the question/answer context. Finally, the model generates answers by inferring facts in the memory.

Performance and limitations. The evaluation of KDMN is based on Visual7W. The framework automatically generates a number of arbitrary question/answer pairs, and its performance was evaluated on open-domain VQA. KDMN performs well to varying degrees on various questions, such as *who*(5.9%), *what*(4.9%), *when*(1.4%) and *how*(2.0%), likely because *who* and *what* questions have a greater variety of questions and multiple-choice questions than do other types of questions, and the system can benefit more from external knowledge.

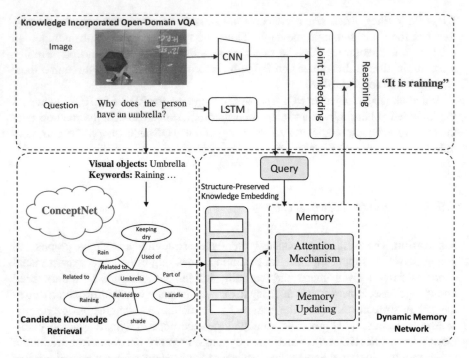

Fig. 5.5 Overall architecture of the memory network query method

References

1. P. Anderson, X. He, C. Buehler, D. Teney, M. Johnson, S. Gould, L. Zhang, Bottom-up and top-down attention for image captioning and visual question answering, in *Proceedings of the IEEE Conference on Computer Vision and Pattern Recognition* (2018), pp. 6077–6086
2. S. Auer, C. Bizer, G. Kobilarov, J. Lehmann, R. Cyganiak, Z.G. Ives, DBpedia: a nucleus for a web of open data, in *The Semantic Web, 6th International Semantic Web Conference, 2nd Asian Semantic Web Conference, ISWC 2007/ASWC 2007, Busan, Korea, 11–15 November 2007*, vol. 4825 (Springer, 2007), pp. 722–735
3. H. Ben-younes, R. Cadène, M. Cord, N. Thome, MUTAN: multimodal tucker fusion for visual question answering, in *Proceedings of the IEEE International Conference on Computer Vision* (IEEE Computer Society, 2017), pp. 2631–2639
4. Q. Cao, B. Li, X. Liang, L. Lin, Explainable high-order visual question reasoning: a new benchmark and knowledge-routed network (2019), arXiv:1909.10128
5. Q. Cao, B. Li, X. Liang, K. Wang, L. Lin, Knowledge-routed visual question reasoning: challenges for deep representation embedding. IEEE Trans. Neural Netw. Learn. Syst. (2021)
6. J. Devlin, M.-W. Chang, K. Lee, K. Toutanova, BERT: pre-training of deep bidirectional transformers for language understanding, in *Proceedings of the Conference of North American Chapter of Association for Computational Linguistics* (2019), pp. 4171–4186
7. F. Gardères, M. Ziaeefard, B. Abeloos, F. Lécué, ConceptBert: concept-aware representation for visual question answering, in *Proceedings of the Conference on Empirical Methods in Natural Language Processing*, ed. by T. Cohn, Y. He, Y. Liu (Association for Computational Linguistics, 2020), pp. 489–498

8. Y. Goyal, T. Khot, D. Summers-Stay, D. Batra, D. Parikh, Making the V in VQA matter: elevating the role of image understanding in visual question answering, in *Proceedings of the IEEE Conference on Computer Vision and Pattern Recognition* (2017), pp. 6325–6334

9. J. Kim, J. Jun, B. Zhang, Bilinear attention networks, in *Proceedings. Advances in Neural Information Processing System*, ed. by S. Bengio, H.M. Wallach, H. Larochelle, K. Grauman, N. Cesa-Bianchi, R. Garnett (2018), pp. 1571–1581

10. R. Krishna, Y. Zhu, O. Groth, J. Johnson, K. Hata, J. Kravitz, S. Chen, Y. Kalantidis, L. Li, D.A. Shamma, M.S. Bernstein, L. Fei-Fei, Visual genome: connecting language and vision using crowdsourced dense image annotations. Int. J. Comput. Vis

11. G. Li, H. Su, W. Zhu, Incorporating external knowledge to answer open-domain visual questions with dynamic memory networks. CoRR (2017), arXiv:1712.00733

12. H. Liu, P. Singh, ConceptNet—a practical commonsense reasoning tool-kit. BT Technol. J. **22**, 211–226 (2004)

13. K. Marino, M. Rastegari, A. Farhadi, R. Mottaghi, OK-VQA: a visual question answering benchmark requiring external knowledge, in *Proceedings of the IEEE Conference on Computer Vision and Pattern Recognition* (2019), pp. 3195–3204

14. M. Narasimhan, A.G. Schwing, Straight to the facts: learning knowledge base retrieval for factual visual question answering, in *Proceedings of the European Conference on Computer Vision*, vol. 11212, ed. by V. Ferrari, M. Hebert, C. Sminchisescu, Y. Weiss (Springer, 2018), pp. 460–477

15. M. Narasimhan, S. Lazebnik, A.G. Schwing, Out of the box: reasoning with graph convolution nets for factual visual question answering, in *Proceedings. Advances in Neural Information Processing Systems*, ed. by S. Bengio, H.M. Wallach, H. Larochelle, K. Grauman, N. Cesa-Bianchi, R. Garnett (2018), pp. 2659–2670

16. E. Perez, F. Strub, H. de Vries, V. Dumoulin, A.C. Courville, FILM: visual reasoning with a general conditioning layer, in *Proceedings of the Conference on AAAI* (2018), pp. 3942–3951

17. S. Ren, K. He, R.B. Girshick, J. Sun, Faster R-CNN: towards real-time object detection with region proposal networks. Proc. Adv. Neural Inf. Process. Syst. **39**, 1137–1149 (2015)

18. V. Shevchenko, D. Teney, A.R. Dick, A. van den Hengel, Reasoning over vision and language: exploring the benefits of supplemental knowledge (2021), ArXiv

19. R. Speer, J. Chin, C. Havasi, ConceptNet 5.5: an open multilingual graph of general knowledge, in *Proceedings of the Conference on AAAI* (2017), pp. 4444–4451

20. H. Tan, M. Bansal, LXMERT: learning cross-modality encoder representations from transformers, in *Proceedings of the Conference on Empirical Methods in Natural Language Processing*, ed. by K. Inui, J. Jiang, V. Ng, X. Wan (Association for Computational Linguistics, 2019), pp. 5099–5110

21. N. Tandon, G. de Melo, F.M. Suchanek, G. Weikum, WebChild: harvesting and organizing commonsense knowledge from the web, in *Seventh ACM International Conference on Web Search and Data Mining, WSDM 2014, New York, NY, USA, 24–28 February 2014* (ACM, 2014), pp. 523–532

22. N. Tandon, G. de Melo, G. Weikum, Acquiring comparative commonsense knowledge from the web, in *Proceedings of the Conference on AAAI* (2014), pp. 166–172

23. P. Wang, Q. Wu, C. Shen, A.R. Dick, A. van den Hengel, Explicit knowledge-based reasoning for visual question answering, in *Proceedings of the International Joint Conference on Artificial Intelligence*, ed. by C. Sierra (2017), pp. 1290–1296

24. P. Wang, Q. Wu, C. Shen, A.R. Dick, A. van den Hengel, FVQA: fact-based visual question answering. IEEE Trans. Pattern Anal. Mach. Intell. **40**(10), 2413–2427 (2018)

25. Q. Wu, P. Wang, C. Shen, A.R. Dick, A. van den Hengel, Ask me anything: free-form visual question answering based on knowledge from external sources, in *Proceedings of the IEEE Conference on Computer Vision and Pattern Recognition* (2016), pp. 4622–4630

26. J. Yu, Z. Zhu, Y. Wang, W. Zhang, Y. Hu, J. Tan, Cross-modal knowledge reasoning for knowledge-based visual question answering. Pattern Recognit. **108**, 107563 (2020)

27. Z. Yu, J. Yu, C. Xiang, J. Fan, D. Tao, Beyond bilinear: generalized multimodal factorized high-order pooling for visual question answering. IEEE Trans. Neural Netw. Learn. Syst. **29**, 5947–5959 (2018)

28. Z. Yu, J. Yu, Y. Cui, D. Tao, Q. Tian, Deep modular co-attention networks for visual question answering, in *Proceedings of the IEEE Conference on Computer Vision and Pattern Recognition* (2019), pp. 6274–6283
29. Z. Zhu, J. Yu, Y. Wang, Y. Sun, Y. Hu, Q. Wu, Mucko: multi-layer cross-modal knowledge reasoning for fact-based visual question answering, in *Proceedings of the International Joint Conference on Artificial Intelligence*, ed. by C. Bessiere (2020), pp. 1097–1103. ijcai.org

Chapter 6
Vision-and-Language Pretraining for VQA

Abstract Multimodal (e.g., vision and language) pretraining has emerged as a popular topic, and many representation learning models have been proposed in recent years. In this chapter, we focus on the vision-and-language pretraining model, which can be adapted in the VQA task. To this end, we first introduce three general pretraining models—ELMo, GPT and BERT—for which only the representation of natural language is considered in the original research. Subsequently, we describe the vision-and-language pretraining models, which can be regarded as extensions of the language-aware pretraining models. Specifically, we categorize these models into two types: single stream and two stream. Finally, we describe the method to finetune these models for each specific downstream task, e.g., VQA, visual common-sense reasoning (VCR) and referring expression comprehension (REC).

6.1 Introduction

Vision-and-language pretraining has attracted considerable attention in recent years. This process is aimed at learning a task-agnostic joint representation of both visual content (e.g., images and videos) and natural language. To this end, the model must understand visual concepts, language semantics and alignment and the relationship between these two modalities. Therefore, many researchers [1, 3, 8, 9, 11, 12, 19, 20] have attempted to develop more promising joint representations. Visual question answering (VQA) is a key downstream task of vision-and-language pretraining methods. In this chapter, we focus on methods to adapt well-designed pretraining models to the VQA task. To this end, we first introduce three typical pretraining models (i.e., ELMo [15], GPT [17] and BERT [4]), which consider only natural language in their original versions, in Sect. 6.2. Subsequently, we present a series of vision-and-language extensions (vision-and-language pretraining models) in Sect. 6.3, which can be categorized into two types: single stream and two stream. In the single-stream models, visual information and linguistic information are fused at the beginning and directly and concurrently input to the encoder (transformer) module. In the other models (i.e., two streams), the visual and linguistic information first passes through two independent encoder (transformer) modules, and a cross transformer is used to

Q. Wu et al., *Visual Question Answering*, Advances in Computer Vision and Pattern Recognition, https://doi.org/10.1007/978-981-19-0964-1_6

integrate the output from different modalities. Finally, in Sect. 6.4, we discuss the methods to adapt the vision-and-language pretrained models in the VQA task and other downstream tasks.

6.2 General Pretraining Models

In this section, we introduce several general pretraining models designed for generating a general representation of a given input. We consider three classical pretraining models (i.e., ELMo [15], GPT [17] and BERT [4]) as examples, which are based on deep neural networks (DNNs). The original versions of these three models consider only natural language. Following the settings of the original versions, we also consider natural language as the input when describing the models.

6.2.1 Embeddings from Language Models

The main idea of embeddings from language models (ELMo) [15] is to optimize a language model on a large amount of unlabeled data via a deep bidirectional RNN (BiLSTM) (as shown in Fig. 6.1). In this context, this language model processes the input sentence and obtains the output vector, which can be regarded as a feature extractor. Unlike the pretraining of Word2Vec [13] or GloVe [14], the embedding obtained by ELMo is contextual due to the BiLSTM, which enables the ELMo model to learn the context information from a given sentence (i.e., a series of word tokens).

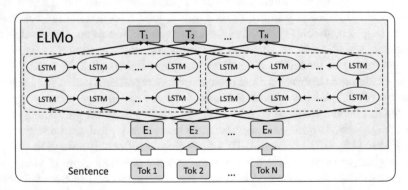

Fig. 6.1 Overall architecture of embeddings from language models (ELMos). The input sentence is represented by a series of word tokens, which are fed to the ELMo model. The output is the corresponding representation for each input token

6.2.2 Generative Pretraining Model

According to the framework shown in Fig. 6.2a, Radford et al. [17] proposed a generative pretraining (GPT) model, which aims to learn a general representation of natural language. To this end, the GPT model seeks to capture long-term dependencies in sentences, replacing the conventional LSTM with a series of transformer [21] modules. Similar to the ELMo model, the GPT model takes a set of word tokens (i.e., sentence) as input and outputs the corresponding representation for each input token. However, as the GPT model is a single directional model, it predicts the current word based on only the previous words, thereby limiting the capability of understanding the context of an input sentence. To alleviate this issue, a bidirectional encoder representations from transformers (BERT) model has been proposed, which focuses on the forward and backward directions of the input sentence (Fig. 6.2b). Notably, the BERT model is widely used in many existing vision-and-language pretraining models [9, 11, 19]. More details regarding the model are presented in the following text.

6.2.3 Bidirectional Encoder Representations from Transformers

As shown in Figs. 6.2b and 6.3, Devlin et al. [4] devised a language representation model known as bidirectional encoder representations from transformers (BERT), which learns deep bidirectional representations from unlabeled sentences by jointly using conditions on both left and right contexts in all layers.

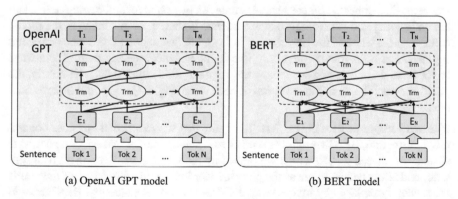

(a) OpenAI GPT model   (b) BERT model

Fig. 6.2 Overall architectures of the **a** OpenAI generative pretraining (GPT) model and **b** bidirectional encoder representations from transformers (BERT) model. The notation "Trm" denotes the transformer module. The inputs are a series of word tokens, while the outputs are the corresponding representations of the input tokens

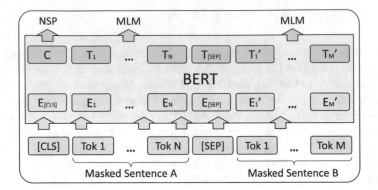

Fig. 6.3 Overall BERT architecture. The input is an unlabeled sentence pair of A and B. The [CLS] symbol is added in front of every input example, while [SEP] represents a separator token between two sentences. NSP refers to a next sentence prediction model, which aims to distinguish whether the input two sentences are relevant. The masked language model (MLM) seeks to learn a deep bidirectional representation

Input Representation. The inputs of BERT are shown in Fig. 6.4. Two sentences, "*my dog is cute*" and "*he likes playing*", are fed into the model. Special tokens [CLS] and [SEP] are added at the beginning and end of the first sentence separately, which indicate the start and end of the first sentence, respectively. For the second sentence, an ending token [SEP] is placed after "*##ing*" (i.e., the end of the second sentence). This method divides "*playing*" into two tokens, "*play*" and "*##ing*". In this manner, the model can manage a word that it has never encountered before, like "*playing*". Next, each input word is represented by three embeddings: a token embedding, a segment embedding and a position embedding. Each token embedding refers to a word represented as a feature vector in common space. As the input only involves two segments (i.e., either the first or second sentence), two segment embeddings exist, and the embedding derived from the same sentence is shared. In this manner, the model can recognize whether the information is from the first segment/sentence or second segment. For tasks that contain only one input sentence, the segment id is always 0; for tasks that input two sentences, the segment id is either 0 or 1. Similarly, the position embedding maps the position of each word (Fig. 6.4) into a low-dimensional dense vector.

Masked Language Model (MLM). To ensure that the model can consider contextual information, BERT introduces a masked language model (MLM). The MLM is similar to a closed test, in which we randomly hide a word involved in a given sentence and allow people to guess the possible words. Specifically, 15%[1] of the words are randomly replaced by tokens named [Mask], and subsequently, BERT tries to predict the words of these [Mask] tokens. The probability of correct prediction is maximized by optimizing the model parameters by using the cross-entropy loss.

---

[1] Excessively little masking impedes the learning of the information from context, while excessively large masking can increase the computational cost.

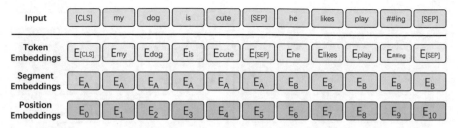

Fig. 6.4 Input representations of BERT. We consider two sentences as inputs, e.g., "*my dog is cute*" and "*he likes playing*". The input representation of each word contains three embeddings: a token embedding, a segment embedding and a position embedding. The token embedding is a dense vector of words. The segment embedding indicates whether the current word is from the first or second sentence. The position embedding denotes the order of each word in the two input sentences

This approach enforces the BERT model to consider contextual information when encoding a word.

However, a problem remains: the special token [Mask] appears in the training phase of MLM; however, this token does not appear in the finetuning phase. Thus, the presence of several new words (i.e., words not encountered in the training phase) may lead to a mismatch when finetuning BERT in other downstream tasks. To alleviate this issue, in pretraining BERT, if a token is among the selected 15% of words, it would be randomly executed in the following three ways:

- 80% of the selected tokens are replaced by [Mask], e.g.,
  my dog is hairy → *my dog is [Mask]*
- 10% of the selected tokens are replaced by a random word, e.g.,
  my dog is hairy → *my dog is apple*
- 10% of the selected tokens are not replaced, e.g.,
  my dog is hairy → *my dog is hairy*

In this scenario, the BERT model does not know which word the [Mask] replaces, as any word may be replaced, e.g., the word "*apple*" may be the replaced word. Hence, the model does not only rely on the current word when encoding the current word, but also scan the context. This framework helps fill in the blanks ([Mask]) or correct the replaced/mismatching word in the sentence. For instance, in the above example, when encoding word "*apple*", if the model considers the context "*my dog is*", it would output an embedding of "*hairy*" instead of "*apple*".

Next Sentence Prediction (NSP). In many natural language processing (NLP) tasks, such as question answering (QA), relationships between sequential sentences are vital. Therefore, BERT introduces a new module known as next sentence prediction (NSP), which aims to predict whether two given sentences are related. This framework requires the pretraining data to be an "article", which contains multiple sequential sentences. To this end, the data from the BookCorpus [25] dataset and English Wikipedia are used when optimizing this model. BookCorpus contains books, and the sentences in each book are related. Similarly, the sentences in English Wikipedia

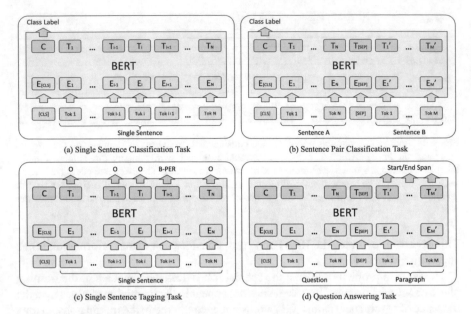

(a) Single Sentence Classification Task   (b) Sentence Pair Classification Task

(c) Single Sentence Tagging Task   (d) Question Answering Task

Fig. 6.5 Downstream BERT tasks. We consider four types of downstream tasks for which the BERT model can be finetuned

are also related. To accomplish this task, BERT selects sequential (related) sentences with a probability of 50% while randomly selecting two unrelated sentences with a probability of 50%. Subsequently, the model determines whether the selected two sentences are related. For example, the following two sentences are related:

[CLS] the man went to [MASK] store [SEP] he bought a gallon [MASK] milk [SEP]

The following sentences are unrelated:

[CLS] the man [MASK] to the store [SEP] penguin [MASK] are flight ##less birds [SEP]

Finetuning to Downstream Tasks. As shown in Fig. 6.5, BERT can be finetuned to accomplish four types of downstream tasks: single sentence classification task, sentence pair classification task, single sentence tagging task and question answering task. The tasks can be described as follows.

- For single sentence classification tasks, the input is a single sentence (Fig. 6.5a), and all tokens belong to the same segment (i.e., id=0). In this task, a softmax function is added to the last layer of the model, and a series of labeled data is used for finetuning.
- For sentence pair classification tasks, as shown in Fig. 6.5b, given two sentences, each token may correspond to different sentences/segments (id=0 and id=1 correspond to the first and second sentences, respectively). Moreover, the model introduces a softmax function in the last layer, and the modified model uses labeled data for finetuning.
- For single sentence tagging tasks (e.g., named entity recognition, NER), given an input sentence (i.e., token sequence), there exists an output tag for each input token

except for [CLS] and [SEP]. For example, as shown in Fig. 6.5c, the notation "B-PER" represents the beginning of a person's name, while "O" indicates that a token belongs to no entity. In this sense, the model is finetuned by evaluating the differences between the predicted tags and ground-truth tags.

- For the question answering task, the input is a question (Q) and a long paragraph (P) containing answers, while the output is the answers (A) found in this paragraph. E.g.,

Q: *Where do water droplets collide with ice crystals to form precipitation?*
P: *... Precipitation forms as smaller droplets coalesce via collision with other rain drops or ice crystals within a cloud ...*
A: *within a cloud*

The BERT mechanism for this task can be summarized as follows: The model first represents the question and paragraph as a long sequence, separated by [SEP] in the middle. The question is a segment/sentence (id=0), while the paragraph containing the answer is another segment/sentence (id=1). It is assumed that the answer is a continuous sequence (i.e., span in Fig. 6.5d) in the paragraph. BERT transforms the problem of finding an answer into a problem of finding the starting and ending indexes of this span.

6.3   Commonly Used Methods for Vision-and-Language Pretraining

Problem Definition and Pretraining Paradigms. Vision-and-language pretraining aims to produce a joint representation from both visual and linguistic inputs. Specifically, given an image (vision) with the corresponding description (language), the model seeks to yield a uniform representation, which retains both text context and visual information. This representation can be applied to a variety of vision-and-language downstream tasks, such as visual question answering (VQA) or referring expression comprehension (REC). To this end, many paradigms for vision-and-language pretraining frameworks have been proposed, which can be categorized into two types: single stream and two stream (Fig. 6.6). Notably, the two-stream paradigm can be further divided into two subtypes, i.e., cross type and joint type.

Notations. Based on BERT [4], each token/word is projected to a corresponding embedding $e \in \mathcal{E}$, where $\mathcal{E}$ denotes the set of word embeddings. Each word embedding e consists of three parts: a token embedding e_{tok}, segment embedding e_{seg} and position embedding e_{pos}. For the input image, similar to the processing of the word embedding, the model considers each bounding region in the image as an input token, which can be mapped to a visual embedding $f \in \mathcal{F}$. Here, $\mathcal{F}$ refers to the set of visual embeddings. Similar to the word embedding, each visual embedding contains three components, i.e., a visual feature representation f_{vis}, a segment embedding f_{seg} and a position embedding f_{pos}. The visual feature representation f_{vis} is the feature of the bounding region, often generated by a convolutional neural network (CNN). The

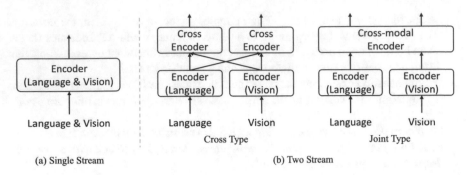

Fig. 6.6 Various paradigms for vision-and-language pretraining frameworks. The methods can be categorized into two types: single stream and two stream. Two-stream frameworks involve two subtypes: cross type and joint type

segment embedding $\mathbf{f}_{seg}$ aims to indicate whether this embedding is derived from the input image or input text.

6.3.1   Single-Stream Methods

Motivation. Many vision-and-language tasks require the understanding of visual contents, linguistic semantics, cross-modal alignments and relationships. A straightforward method is to use separate vision-and-language models designed for vision or language tasks that are pretrained on vision or language datasets, respectively. However, this method lacks a unified foundation for learning joint representations among visual concepts and linguistic semantics. Consequently, the method often exhibits an inferior generalization ability when the paired vision-and-language data are limited or biased. To alleviate this issue, a jointly pretrained vision-and-language model must be considered, which can provide a joint knowledge representation for downstream vision-and-language tasks.

Methods. Li et al. [9] proposed a vision-language representation framework known as VisualBERT (Fig. 6.7), aimed at producing a unified representation, which contains both linguistic semantics (from the caption) and visual concepts (from the image). Specifically, to capture rich semantics in images and corresponding text, VisualBERT employs BERT [4] (a transformer-based model [21]) for natural language processing and Faster R-CNN [18] (an object detection model) to generate region proposals from images. Similar to the processing of tokens in BERT, Visual-BERT takes each bounding region as an input token and feeds it to the model along with the word tokens. Each bounding region is mapped to a visual feature via a CNN model. Subsequently, the text and image features are jointly processed by multiple transformer layers. The interaction among words and object regions empowers the model to consider the associations between text and image. VisualBERT is trained

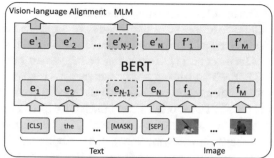

The man at bat readies to swing at the pitch while the umpire looks on

Fig. 6.7 Overall architecture of VisualBERT (a typical single-stream vision-and-language pretraining framework). The model takes both text and image as inputs and seeks to produce a joint representation that contains both linguistic semantics and visual concepts

on the COCO dataset [2] with two objectives. The first objective is masked language modeling (MLM), following the settings in BERT. However, unlike BERT based only on natural language, VisualBERT considers visual information when filling the [MASK] token. The second objective is sentence-image prediction, which focuses on whether or not the given text and image are aligned. To this end, the model takes an image with two captions as input. One caption is associated with the given image, while the other caption only has a 50% chance of being relevant.

Li et al. [8] devised a universal encoder named Unicoder-VL, which seeks to learn joint representations of vision and language in a pretraining manner. Specifically, inspired by cross-lingual pretrained models, e.g., XLM [7], Unicoder [6] and BERT [4], Unicoder-VL uses a multi-layer transformer [21] to capture both visual and linguistic contents from the multimodal inputs. The optimization of Unicoder-VL involves three objectives, including masked language modeling (MLM), masked object classification (MOC) and visual-linguistic matching (VLM). Similar to VisualBERT, MLM forces the model to consider both linguistic semantics from text and visual information from images. In addition to masking the text tokens, MOC enhances the context-aware representation ability of the generated multimodal features by masking the detected objects. Finally, VLM tries to predict whether an image and a text describe each other, similar to the VisualBERT framework.

Su et al. [19] proposed a new model for visual-linguistic tasks, named visual-linguistic BERT (VL-BERT). The VL-BERT model adopts a multi-layer transformer as the backbone, which takes the text tokens and image regions as inputs and outputs features with both visual and linguistic information. The model is optimized via two pretrained tasks: masked language modeling with visual clues and masked RoI classification with linguistic clues. The former task is similar to the masked language modeling (MLM) implemented in BERT. The difference is that VL-BERT focuses on both linguistic and visual clues from the input text and image, respectively, while the conventional BERT model considers only natural language. Unlike the first task, which seeks to mask the tokens in the input sentence, the second task aims to predict the masked region of interest (RoI) in a given image. In contrast to the Unicoder-VL or VisualBERT framework, VL-BERT masks the RoIs at the pixel level, while other methods add masks in the features of RoIs. In this manner, VL-BERT can avoid visual clue leakage during the process of visual feature extraction.

Alberti et al. [1] designed a model named bounding boxes in text transformer (B2T2), which tries to verify whether a satisfactory integration of visual and linguistic information can enhance the model performance in downstream tasks. To this end, B2T2 introduces two kinds of fusion methods: late fusion and early fusion. Inspired by dual encoder models [5, 22], which map a whole image into a common representation space, the late fusion method maps both the input image and sentence into a common space at the end of the backbone model and calculates their inner production as an output score. In contrast to late fusion, the early fusion method seeks to integrate the image and sentence at the beginning of the backbone model. To this end, this model utilizes the [MASK] mechanism similar to BERT, which also adopts the [MASK] token into object regions of the given image. Model training is based on two pretraining tasks, i.e., masked language model (MLM) and sentence-image alignment, which are the same as those of VisualBERT.

Chen et al. [3] devised a universal image-text representation (UNITER) model. Unlike the existing approaches, which consist of a multi-layer transformer, UNITER leverages only a single transformer. The input image is configured following the settings of existing approaches (e.g., VisualBERT), which rely on the region of interest (RoI) that contains the visual features and position information. The key difference is that the position is represented by a vector with 7 dimensions (i.e., height, width, and area of the bounding box), while other methods consist of only 4 dimensions (i.e., coordinates of the bounding box). The optimization of this model is based on four pretraining tasks: masked language modeling (MLM), masked region modeling (MRM), image-text matching (ITM) and word-region alignment (WRA). The first three tasks (i.e., MLM, MRM and ITM) are the same as those in Unicoder-VL. In addition to ITM (global image-text alignment), UNITER introduces a new WRA, which focuses on fine-grained alignment between words and image regions. Specifically, this model regards the problem of token/word-region matching as a transfer problem between two distributions. In this sense, the model can use the optimal transportation (OT) approach to encourage the alignment between words and regions in an unsupervised manner. The corresponding loss estimates the OT distance between the word and region distributions.

Performance and limitations. In addition to the abovementioned methods, many variants, like VLP [24], ImageBERT [16], XGPT [23] and OSCAR [10], have been established. However, due to the limitation of the model structure, the single-stream models cannot accommodate the different processing requirements associated with different modalities (i.e., vision and language), which limits the interaction among different modalities, as the interaction must be able to occur in varying representation depths. Such a rigid architecture may hamper the general ability of the yielded multimodal features and further deteriorate the finetuning performance in downstream tasks.

6.3.2 Two-Stream Methods

Motivation. For two-stream models, the linguistic semantics and visual information are not directly fused at the beginning of the model but are first encoded by different encoders. The shunt design is based on the assumption that language understanding is more complex than image understanding, and the image input is a series of higher-level features extracted by the object detection model (e.g., Faster R-CNN). Thus, the encoding required by the two inputs should be different (e.g., different model or representation depths).

Methods. Lu et al. [11] proposed a model known as vision-and-language BERT (ViLBERT) for learning the joint representations of image content and natural language. As shown in Fig. 6.8, the model consists of two streams, which first separately process the visual and textual inputs, and the two streams interact with each other via coattention transformer layers. The optimization of ViLBERT is based on two pretraining objectives: masked multimodal modeling and multimodal alignment prediction. Masked multimodal modeling follows the MLM of conventional BERT, which masks both the words/tokens and image regions in the input and tries to recover them in the output of this model. Notably, rather than recovering the masked feature value for each image region, ViLBERT seeks to predict the distribution of semantic classes for each region. To this end, the model considers the predicted distribution as the ground truth, which is derived from the same detector used for region detection. Subsequently, the model ensures that the recovered distribution is similar to the predicted distribution via KL diversity. Multimodal alignment prediction focuses on whether or not the given image and sentence are aligned.

Tan et al. [20] devised a framework for learning cross-modality encoder representations from transformers (LXMERT), which seeks to learn the vision-and-language connections from input images and sentences. The model consists of three components: an object relationship encoder, a language encoder, and a cross-modality encoder, which focus on capturing visual features, extracting linguistic embeddings and fusing the visual and linguistic information, respectively. LXMERT involves five pretraining tasks, which can be categorized into three types: language

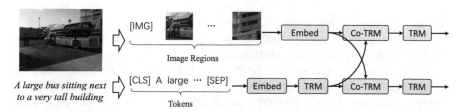

Fig. 6.8 Overall architecture of ViLBERT (a typical two-stream vision-and-language pretraining framework). Given an image with a caption, the model first separately manages the two inputs by a series of transformer blocks (TRM) and realizes interaction through a coattentional TRM (Co-TRM) module

tasks, vision tasks and cross-modality tasks. For language tasks, the model follows masked language modeling (MLM) in BERT. The vision task contains two subtasks: masked object prediction via RoI-feature regression and masked object prediction via detected-label classification. In both tasks, RoI features are randomly masked with zeros. The difference is that the former approach seeks to recover the masked region, while the latter focuses on classifying the given RoI region. Cross-modality tasks also involve two subtasks, i.e., cross-modality matching and visual question answering (VQA). The first task checks whether images and text are aligned, while the second task is the conventional VQA task, which is used to enlarge the pretraining dataset.

Lu et al. [12] designed a multitask vision-and-language representation learning method known as 12-in-1. This method optimizes a model on 12 datasets pertaining to four types of tasks: visual question answering, caption-based image retrieval, grounding referring expressions and multimodal verification. The backbone architecture follows the ViLBERT model, containing two streams pertaining to vision and language. Similar to ViLBERT, 12-in-1 also considers two pretraining tasks, i.e., masked multimodal modeling and multimodal alignment prediction. In the first task, 12-in-1 masks words/tokens and image regions with a probability of approximately 15%. However, unlike the original ViLBERT, 12-in-1 also masks the regions with an overlap (i.e., >0.4 IoU), which can avoid the leakage of visual information. The section task distinguishes whether text and image are matched. In contrast to ViLBERT, 12-in-1 does not implement the masked multimodal modeling loss when processing negative (not aligned) text. In this manner, the model can remove the noise derived from the negative samples to a certain extent.

Performance and limitations. Compared to one-stream architectures, although the two-stream architectures are more flexible when handling visual and linguistic inputs, the models often contain additional parameters with a higher computational cost (e.g., the pretraining process of LXMERT requires 10 d on 4 Titan Xp). It remains a challenge to decrease the computational cost and design a lighter model.

6.4   Finetuning on VQA and Other Downstream Tasks

Pretrained visual-linguistic models can be used in many types of vision-and-language applications, e.g., visual question answering (VQA), visual common-sense reasoning (VCR), referring expression comprehension (REC), natural language visual reasoning, Flick30k entities, image-text retrieval, zero-shot image-text retrieval, grounding referring expressions, visual entailment, image captioning and GQA. Each application/task corresponds to a specific finetuning method, which adapts the original model to a certain extent. In this section, we describe the first three typical downstream tasks.

Visual Question Answering. Visual question answering (VQA) is a vital downstream task that has been attempted to be realized using various vision-and-language

Fig. 6.9 Finetuning on the visual question answering (VQA) task

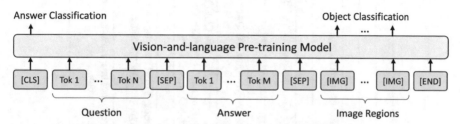

Fig. 6.10 Finetuning on the visual common-sense reasoning (VCR) task

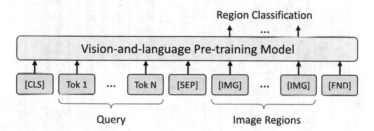

Fig. 6.11 Finetuning on the referring expression comprehension (REC) task

pretraining models, e.g., VisualBERT, VL-BERT, UNITER, ViLBERT, LXMERT and 12-in-1. In general, the VQA task takes an image with an image-dependent question as the input and requires the model to return a proper answer for such a question. As shown in Fig. 6.9, when using a vision-and-language pretraining model (c.g., VisualBERT), VQA can be regarded as a classification problem, in which the model must select an appropriate answer from a predefined answer pool. To ensure that the pretrained model is suitable for the VQA task, an additional [MASK] token is introduced, which was fed to the model to return a predicted answer.

Visual common-sense Reasoning. Another downstream task is visual common-sense reasoning (VCR). Similar to VQA, VCR requires the model to generate a correct answer based on the given image and question. The key difference is that in addition to the output answer, VCR necessitates the verification of the reasonability of the generated answer. In this sense, the VCR task can be divided into two subtasks: question (Q) → answer (A) and question (Q) & answer (A) → reasoning (R). Each

Table 6.1 Comparison of different vision-and-language pretraining methods

Method	Architecture	Visual token	Pretraining datasets
VisualBERT [9]	Single stream	Image RoI	COCO
Unicoder-VL [8]	Single stream	Image RoI	Conceptual captions
VL-BERT [19]	Single stream	Image RoI	Conceptual captions, BooksCorpus, English Wikipedia
B2T2 [1]	Single stream	Image RoI/Entire image	Conceptual captions
UNITER [3]	Single stream	Image RoI	COCO, Visual genome, Conceptual captions, SBU captions
ViLBERT [11]	Two stream	Image RoI	Conceptual captions
LXMERT [20]	Two stream	Image RoI	COCO, Visual genome (VG) caption, VG QA, VQAv2, GQA
12-in-1 [12]	Two stream	Image RoI	VQAv2, GQA, Visual genome (VG) QA, COCO, Flickr30K, RefCOCO, RefCOCO+, RefCOCOg, Visual7W, GuessWhat, NLVR, SNLI-VE
Method	Pretraining tasks		Downstream tasks
VisualBERT [9]	Masked language modeling (MLM), Sentence-image alignment		Visual question answering, Visual common-sense reasoning, Natural language visual reasoning, Grounding phrases
Unicoder-VL [8]	Masked language modeling (MLM), Masked object classification (MOC), Visual-linguistic matching (VLM)		Image-text retrieval, Zero-shot image-text retrieval, Visual common-sense reasoning

(continued)

Table 6.1 (continued)

Method	Pretraining tasks	Downstream tasks
VL-BERT [19]	Masked language modeling (MLM), Masked object classification (MOC)	Visual common-sense reasoning, Visual question answering, Referring expression comprehension
B2T2 [1]	Masked language modeling (MLM), Sentence-image alignment	Visual common-sense reasoning
UNITER [3]	Masked language modeling (MLM), Masked region modeling (MRM), ImageText matching (ITM), Word-region alignment (WRA)	Visual question answering, Image-text retrieval, Referring expression comprehension, Visual common-sense reasoning, Visual entailment, NLVR
ViLBERT [11]	Masked language modeling (MLM), Sentence-image alignment, Masked object classification (MOC)	Visual question answering, Visual common-sense reasoning, Grounding referring expressions, Caption-based image retrieval, Zero-shot caption-based image retrieval
LXMERT [20]	Masked language modeling (MLM), Sentence-image alignment, (((((Masked region classification, Masked region feature regression, Visual question answering	Visual question answering, GQA, NLVR
12-in-1 [12]	Masked language modeling (MLM), Sentence-image alignment, Masked object classification (MOC)	Visual question answering, Caption-based image retrieval, Grounding referring expressions, Multimodal verification

training sample contains four candidate answers, which are combined with the given question and images separately. Consequently, there are four combined sequences for each sample. As shown in Fig. 6.10, the model takes sequences as inputs and seeks to classify the input sequence that is correct. Object classification is often introduced as an auxiliary module.

Referring Expression Comprehension. Referring expression comprehension (REC) aims to localize a specific object in a given image, which is described by a linguistic query. According to Fig. 6.11, the task takes a query-image pair as the input and outputs a detected region, which covers the described object of the given query. A vision-and-language pretraining model can be easily adapted in this task, as the inputs contain a linguistic sequence and a series of image regions in both the pretraining and REC tasks. The only difference is the integration of a region classification module in the output layer of the original model (Table 6.1).

References

1. C. Alberti, J. Ling, M. Collins, D. Reitter, Fusion of detected objects in text for visual question answering, in *Proceedings of the Conference on Empirical Methods in Natural Language Processing* (2019)
2. X. Chen, H. Fang, T.-Y. Lin, R. Vedantam, S. Gupta, P. Dollár, C.L. Zitnick, Microsoft COCO captions: data collection and evaluation server (2015), arXiv:1504.00325
3. Y.-C. Chen, L. Li, L. Yu, A. El Kholy, F. Ahmed, Z. Gan, Y. Cheng, J. Liu, UNITER: universal image-text representation learning, in *Proceedings of the European Conference on Computer Vision* (Springer, 2020), pp. 104–120
4. J. Devlin, M.-W. Chang, K. Lee, K. Toutanova, BERT: pre-training of deep bidirectional transformers for language understanding, in *NAACL-HLT* (2019)
5. D. Gillick, A. Presta, G.S. Tomar, End-to-end retrieval in continuous space (2018), arXiv:1811.08008
6. H. Huang, Y. Liang, N. Duan, M. Gong, L. Shou, D. Jiang, M. Zhou, Unicoder: a universal language encoder by pre-training with multiple cross-lingual tasks (2019), arXiv:1909.00964
7. G. Lample, A. Conneau, Cross-lingual language model pretraining (2019), arXiv:1901.07291
8. G. Li, N. Duan, Y. Fang, M. Gong, D. Jiang, Unicoder-VL: a universal encoder for vision and language by cross-modal pre-training, in *Proceedings of the Conference on AAAI*, vol. 34 (2020), pp. 11336–11344
9. L.H. Li, M. Yatskar, D. Yin, C.-J. Hsieh, K.-W. Chang, VisualBERT: a simple and performant baseline for vision and language (2019), arXiv:1908.03557
10. X. Li, X. Yin, C. Li, P. Zhang, X. Hu, L. Zhang, L. Wang, H. Hu, L. Dong, F. Wei et al., Oscar: object-semantics aligned pre-training for vision-language tasks, in *Proceedings of the European Conference on Computer Vision* (Springer, 2020)
11. J. Lu, D. Batra, D. Parikh, S. Lee, ViLBERT: pretraining task-agnostic visiolinguistic representations for vision-and-language tasks, in *Proceedings. Advances in Neural Information Processing Systems* (2019)
12. J. Lu, V. Goswami, M. Rohrbach, D. Parikh, S. Lee, 12-in-1: multi-task vision and language representation learning, in *Proceedings of the IEEE Conference on Computer Vision and Pattern Recognition* (2020), pp. 10437–10446
13. T. Mikolov, K. Chen, G. Corrado, J. Dean, Efficient estimation of word representations in vector space (2013), arXiv:1301.3781

14. J. Pennington, R. Socher, C.D. Manning, Glove: global vectors for word representation, in *Proceedings of the Conference on Empirical Methods in Natural Language Processing* (2014), pp. 1532–1543
15. M.E. Peters, M. Neumann, M. Iyyer, M. Gardner, C. Clark, K. Lee, L. Zettlemoyer, Deep contextualized word representations (2018), arXiv:1802.05365
16. D. Qi, L. Su, J. Song, E. Cui, T. Bharti, A. Sacheti, ImageBERT: cross-modal pre-training with large-scale weak-supervised image-text data (2020), arXiv:2001.07966
17. A. Radford, K. Narasimhan, T. Salimans, I. Sutskever, Improving language understanding by generative pre-training (2018)
18. S. Ren, K. He, R. Girshick, J. Sun, Faster R-CNN: towards real-time object detection with region proposal networks, in *Proceedings. Advances in Neural Information Processing Systems* (2015)
19. W. Su, X. Zhu, Y. Cao, B. Li, L. Lu, F. Wei, J. Dai, VL-BERT: pre-training of generic visual-linguistic representations, in *Proceedings of the International Conference on Learning Representations* (2020)
20. H. Tan, M. Bansal, LXMERT: learning cross-modality encoder representations from transformers, in *Proceedings of the Conference on Empirical Methods in Natural Language Processing* (2019)
21. A. Vaswani, N. Shazeer, N. Parmar, J. Uszkoreit, L. Jones, A.N. Gomez, L. Kaiser, I. Polosukhin, Attention is all you need, in *Proceedings. Advances in Neural Information Processing Systems* (2017)
22. L. Wu, A. Fisch, S. Chopra, K. Adams, A. Bordes, J. Weston, StarSpace: embed all the things! in *Proceedings of the Conference on AAAI* (2018)
23. Q. Xia, H. Huang, N. Duan, D. Zhang, L. Ji, Z. Sui, E. Cui, T. Bharti, X. Liu, M. Zhou, XGPT: cross-modal generative pre-training for image captioning (2020), arXiv:2003.01473
24. L. Zhou, H. Palangi, L. Zhang, H. Hu, J. Corso, J. Gao, Unified vision-language pre-training for image captioning and VQA, in *Proceedings of the Conference on AAAI*, vol. 34 (2020), pp. 13041–13049
25. Y. Zhu, R. Kiros, R. Zemel, R. Salakhutdinov, R. Urtasun, A. Torralba, S. Fidler, Aligning books and movies: towards story-like visual explanations by watching movies and reading books, in *Proceedings of the IEEE International Conference on Computer Vision* (2015)

Part III
Video-Based VQA

Video question answering takes videos as the input and answers questions regarding video content. In contrast to the image-based VQA problem, video QA requires a model to understand both the spatial and temporal contexts, which naturally exist in a video. In this part, we focus on video QA methodologies. We describe video representation learning and introduce several classical and advanced models for video QA.

Chapter 7
Video Representation Learning

Abstract Video representation learning generates visual semantic representations from given videos, which is vital for video-related tasks, including human action understanding in videos and video question answering. Video representations can be categorized into handcrafted local features and deep-learned features. Handcrafted local features are video features extracted by handcrafted formulas, and deep-learned features are extracted automatically through neural networks. In this chapter, we discuss video representation learning from the two aspects of handcrafted features and deep architecture-generated features.

7.1 Handcrafted Local Video Descriptors

The calculation of handcrafted local video features involves two processing units: (1) a detector to identify the significant and informative regions and (2) a descriptor to generate semantic information regarding the extracted regions. There exist several typical handcrafted local features:

1. Space-time interest points [8]
2. Cuboids [5]
3. Dense trajectories [10]

Space-Time Interest Points The solution of space-time interest points [8] is based on considering the video as a three-dimensional function and identifying a mapping function to map the three-dimensional video to a one-dimensional space, in which the local maximum value is sought. The calculation process can be described as follows: First, we convert the video into a linear scale space representation as

$$L\left(\cdot; \sigma_l^2, \tau_l^2\right) = g\left(\cdot; \sigma_l^2, \tau_l^2\right) * f(\cdot)$$

Here, $g\left(\cdot; \sigma_l^2, \tau_l^2\right)$ is a Gaussian kernel with distinct spatial variance σ_l^2 and temporal variance τ_l^2. This Gaussian kernel is calculated as

$$g\left(x, y, t; \sigma_l^2, \tau_l^2\right) = \frac{\exp\left(-\left(x^2 + y^2\right)/2\sigma_l^2 - t^2/2\tau_l^2\right)}{\sqrt{(2\pi)^3\sigma_l^4\tau_l^2}}$$

Similar to the Harris algorithm, a 3×3 matrix composed of first-order spatial and temporal derivatives averaged with a Gaussian weighting function is established as

$$\mu = g\left(\cdot; \sigma_i^2, \tau_i^2\right) * \begin{pmatrix} L_x^2 & L_xL_y & L_xL_t \\ L_xL_y & L_y^2 & L_yL_t \\ L_xL_t & L_yL_t & L_t^2 \end{pmatrix}$$

Subsequently, we calculate the three eigenvalues of the μ matrix to obtain the expression form of the Harris corner function in the space-time domain by

$$H = \det(\mu) - k \cdot \text{trace}^3(\mu) = \lambda_1\lambda_2\lambda_3 - k\left(\lambda_1 + \lambda_2 + \lambda_3\right)^3$$

By calculating the positive maximum value of H, space-time interest points can be obtained.

Cuboids Cuboid [5] feature extraction involves four processes: feature detection, cuboid generation, cuboid prototype generation and final behavior descriptor generation. The feature detection process finds the local maximum point of the following function:

$$R = (I * g * h_{ev})^2 + (I * g * h_{od})^2$$

Here

$$g(x, y; \sigma) = \frac{1}{2\pi\sigma^2}e^{\frac{-(x^2+y^2)}{2\sigma^2}}$$

$$h_{ev}(t; \tau, \omega) = -\cos(2\pi t\omega)e^{-\frac{t^2}{\tau^2}}$$

$$h_{od}(t; \tau, \omega) = -\sin(2\pi t\omega)e^{-\frac{t^2}{\tau^2}}.$$

After obtaining the feature points, we generate cuboids with feature points as centers. Subsequently, we transform the cuboids, scale the transformed cuboids into vectors and adopt the PCA to reduce the dimension for generating cuboid feature descriptors. Cuboid prototypes are generated by applying the k-means method on the cuboid feature descriptors, which are finally used to generate behavior descriptors.

The most commonly used handcrafted local video descriptors include the histogram of oriented gradient (HOG) [2], histogram of optical flow (HOF) [9] and motion boundary histogram (MBH) [3]. The histogram of oriented gradient (HOG) [1] captures the edge or gradient structure of images. The normalization over the concatenated vector of normalized cell histograms from all the block regions is the final calculation of the HOG descriptor. The advantage of the HOG descriptor is that it can reflect the local shape, which is of significance for human detection in images.

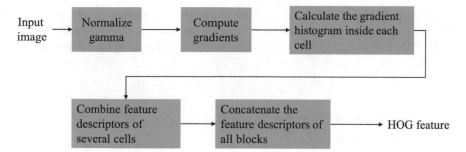

Fig. 7.1 HOG feature extraction

The histogram of oriented optical flow (HOF) descriptor, which is similar to the HOG, performs weighted statistical analyses in the direction of the optical flow to obtain a histogram for the optical flow direction information and is usually used in action recognition (Fig. 7.1).

The motion boundary histogram (MBH) descriptor treats the optical flow images in the x- and y-directions as two grayscale images and extracts the gradient histograms of these grayscale images. Specifically, the MBH feature calculates the HOG feature over the optical flow image in the x- and y-directions of the image. The MBH feature extracts the boundary information of the moving object and can thus be used in pedestrian detection applications. In addition, the calculation of the MBH feature is simple and convenient (Fig. 7.2).

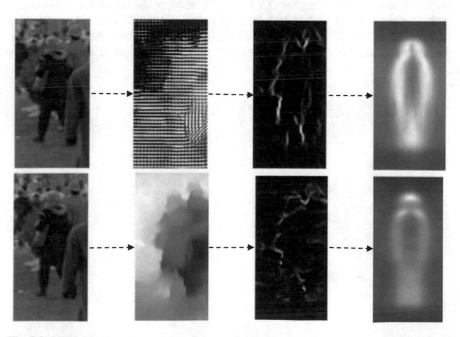

Fig. 7.2 MBH descriptor. From left to right: original image, optical flow, gradient magnitude of flow field and average MBH descriptor

In summary, the abovementioned three local video descriptors are the most commonly used features. HOG is calculated in the image field and is thus a spatial feature, while HOF and MBH are calculated over optical flow images and are thus temporal features.

7.2   Data-Driven Deep Learning Features for Video Representation

Deep learning feature methods aim to automatically learn the semantic representation from raw video by using a deep neural network trained from a large dataset containing labeled data. Compared with handcrafted methods, data-driven deep learning methods are similar to a black box containing a large number of parameters and complex structures. In this section, we introduce several efficient data-driven deep learning features for video representation.

Feichtenhofer et al. [7] proposed a SlowFast network to capture spatiotemporal features. The SlowFast network is a two-stream network, including low and high frame rate streams. Among the retinal ganglion cells of primates, 80 (Fig. 7.3).

Diba et al. [4] proposed temporal 3D ConvNets to generate a video representation and accomplish the video classification task. The temporal transition layer (TTL), which can effectively model the variable time 3D convolution kernels in a shorter or longer range, is used to replace the standard transition layers in DenseNet. Subsequently, a temporal 3D ConvNet (T3D) is built based on the DenseNet architecture, which replaces 2D convolution with 3D convolution and uses TTL to replace the standard transition layers in DenseNet. T3D exhibits excellent performance on a small dataset; however, the number of parameters of the T3D model is 1.3 times larger

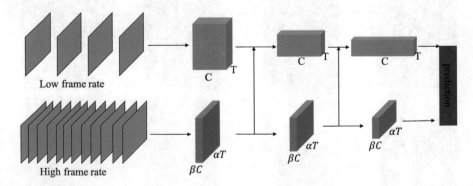

Fig. 7.3 SlowFast network

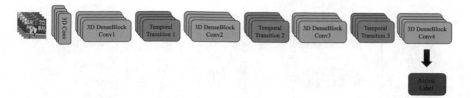

Fig. 7.4 Temporal 3D ConvNet (T3D)

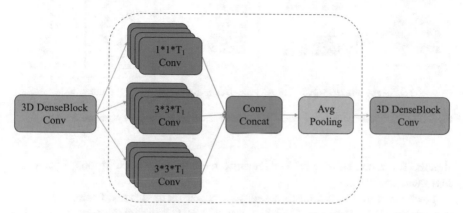

Fig. 7.5 3D temporal transition layer

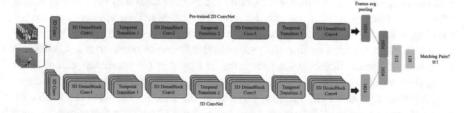

Fig. 7.6 Architecture for knowledge transfer from a pretrained 2D ConvNet to 3D ConvNet

than that for DenseNet3D. Therefore, a transfer learning method is used to transfer the weights of the two-dimensional convolutional network to the three-dimensional convolutional network to decrease the number of parameters (Figs. 7.4, 7.5 and 7.6).

7.3 Self-supervised Learning for Video Representation

Video representation based on supervised learning methods can be expensive. Fine-grained annotations for each frame of the video are required for supervised learning. In addition, training different actions requires new annotations to provide supervision

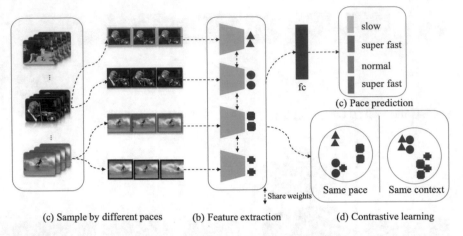

(c) Sample by different paces   (b) Feature extraction   (d) Contrastive learning

Fig. 7.7 Framework for self-supervised video representation learning by pace prediction

signals. Therefore, self-supervised learning methods have been proposed for video representation learning.

Dwibedi et al. [6] proposed a self-supervised method named temporal cycle-consistency (TCC) learning. The key concept of TCC is to find the same action from multiple videos by the principle of cyclic consistency. The goal of the algorithm is to train an effective frame encoder to obtain the representation of the corresponding action. The training process can be described as follows. Two videos are used for training, with one of the videos being the reference video. One frame in the reference video is encoded to find the most similar frame in another video, which is subsequently used to identify the most similar frame in the reference video. If the learned embedding space has cyclic consistency, this frame should be the same frame as the reference frame. The training process of the model decreases the loop consistency error by continuously improving the semantic understanding of each video frame. Since this method can effectively learn the transfer expression of videos, it can be widely used in small-sample video action classification, unsupervised video alignment, multimodal transfer and frame-by-frame video retrieval.

Wang et al. [11] established a new perspective for learning video representations in a self-supervised manner by pace prediction, inspired by the rhythmic montage technique in film production. Training clips are generated using three paces: a slow pace, normal pace and fast pace. Next, the training clips are used to extract spatiotemporal features using a 3D CNN. Later, pace prediction and contrastive learning are simultaneously employed, and the weighted sum of the two losses is the final loss. The main contribution of the paper is that it proposes a new perspective of video representation learning in self-supervised learning, namely, video speed prediction. Moreover, a comparative learning method is used to further regularize the learning process and help the model learn high-level semantic information (Fig. 7.7).

References

1. A.F. Bobick, J.W. Davis, The recognition of human movement using temporal templates. IEEE Trans. Pattern Anal. Mach. Intell. **23**(3), 257–267 (2001)
2. N. Dalal, B. Triggs, Histograms of oriented gradients for human detection, in *2005 IEEE Computer Society Conference on Computer Vision and Pattern Recognition (CVPR'05)*, vol. 1 (IEEE, 2005), pp. 886–893
3. N. Dalal, B. Triggs, C. Schmid, Human detection using oriented histograms of flow and appearance, in *European Conference on Computer Vision* (Springer, 2006), pp. 428–441
4. A. Diba, M. Fayyaz, V. Sharma, A.H. Karami, M.M. Arzani, R. Yousefzadeh, L. Van Gool, Temporal 3d convnets: new architecture and transfer learning for video classification (2017), arXiv:1711.08200
5. P. Dollár, V. Rabaud, G. Cottrell, S. Belongie, Behavior recognition via sparse spatio-temporal features, in *2005 IEEE International Workshop on Visual Surveillance and Performance Evaluation of Tracking and Surveillance* (IEEE, 2005), pp. 65–72
6. D. Dwibedi, Y. Aytar, J. Tompson, P. Sermanet, A. Zisserman, Temporal cycle-consistency learning, in *Proceedings of the IEEE/CVF Conference on Computer Vision and Pattern Recognition* (2019), pp. 1801–1810
7. C. Feichtenhofer, H. Fan, J. Malik, K. He, SlowFast networks for video recognition, in *Proceedings of the IEEE/CVF International Conference on Computer Vision* (2019), pp. 6202–6211
8. I. Laptev, On space-time interest points. Int. J. Comput. Vis. **64**(2), 107–123 (2005)
9. I. Laptev, M. Marszalek, C. Schmid, B. Rozenfeld, Learning realistic human actions from movies, in *2008 IEEE Conference on Computer Vision and Pattern Recognition* (IEEE, 2008), pp. 1–8
10. H. Wang, C. Schmid, Action recognition with improved trajectories, in *Proceedings of the IEEE International Conference on Computer Vision* (2013), pp. 3551–3558
11. J. Wang, J. Jiao, Y.-H. Liu, Self-supervised video representation learning by pace prediction, in *European Conference on Computer Vision* (Springer, 2020), pp. 504–521

Chapter 8
Video Question Answering

Abstract The video question answering task, which was first introduced in 2014, is a more complex task than the classical visual (static image) question answering task. For video question answering tasks, both datasets and models are essential for research. Therefore, in this chapter, we first illustrate the most popular datasets for video question answering, ranging from datasets containing physical objects to those characterizing the real world, and subsequently introduce several models based on the encoder-decoder framework.

8.1 Introduction

The main objective of the video question answering task is to learn a model, for which it is necessary to understand the semantic information in the videos and questions and their semantic correlation to infer the correct answer to the given question [9]. Video question answering can be divided into many subtasks, including video grounding, object detection, feature extraction, multimodal fusion and classification.

The inputs to the model $f(v, q, a; \theta)$ are defined as follows: a video denoted as $v \in V$, a question denoted as $q \in Q$, and an answer denoted as $a \in A$ output by the model. Therefore, the objective function in the learning process is defined as

$$\min_{\theta} L(\theta) = L_\theta + \lambda ||\theta||^2,$$

where θ denotes the model coefficients, L_θ denotes the loss function and λ denotes the trade-off parameter between the training loss and regularization. The method to train model parameters θ to answer the question is the key point to solve the VideoQA task.

Various metrics can be used to evaluate model performance. The evaluation metrics of the general performance, which pertains to answer prediction, are accuracy and WUPS [6]. Temporal mean intersection-over-union [4] is a fine-grained metric to evaluate span prediction (answer-related temporal span). Additionally, the answer-span joint accuracy (ASA) jointly evaluates both answer prediction and span prediction [6].

The remaining section is organized as follows. First, we introduce the most influential video question answering datasets. Subsequently, we describe traditional models for the VideoQA task based on the encoder-decoder framework.

8.2 Datasets

With considerable research efforts devoted to VideoQA, a number of datasets have been established. We broadly classify the existing typical VideoQA datasets based on the complexity and necessary reasoning steps for the questions. Questions in certain datasets require only single-step reasoning, for example, what and how questions. Other datasets pertain to a more complex question like "After walking through a doorway, which object were they interacting with?", which require multistep reasoning. Moreover, according to the video source, datasets can be categorized into movie types, TV types, TGIF types, geometry types, and game and cartoon types. Therefore, in each class, we introduce the datasets in the order of the video source.

8.2.1 Multistep Reasoning Dataset

TVQA [4] is a large-scale compositional video QA dataset based on 6 popular TV shows spanning 3 genres: medical dramas, sitcoms and crime shows. The dataset consists of 152,545 QA pairs from 21,793 clips, spanning over 460 h of video. The video clips in TVQA are relatively long (60?0 s), rendering video understanding challenging. In addition to QA pairs, the dialogue (characters and subtitles) for each video clip is provided. The questions in TVQA are in a compositional-question format: [What/How/Where/Why/...] combined with [when/before/after]. The second part localizes the most relevant frames among the video, and the first part poses a question regarding that relevant frame. One key property of TVQA is that the TVQA dataset provides a temporal timestamp annotation indicating the minimum span (context) needed to answer each question.

TVQA+ [5] is an augmented version of TVQA. While TVQA provides temporal timestamp annotations for each question, it lacks spatial annotations, i.e., bounding boxes of the objects and people. TVQA+ samples one frame every two seconds from each span for spatial annotation and adds framewise bounding boxes for visual concepts mentioned in the questions and correct answers to the dataset. Overall, TVQA+ contains 148,468 images annotated with 310,826 bounding boxes (Figs. 8.1 and 8.2).

SVQA [8] is a large-scale automatically generated dataset. The dataset contains 12,000 videos and 118,680 QA pairs. SVQA generates each video using Unity3D,

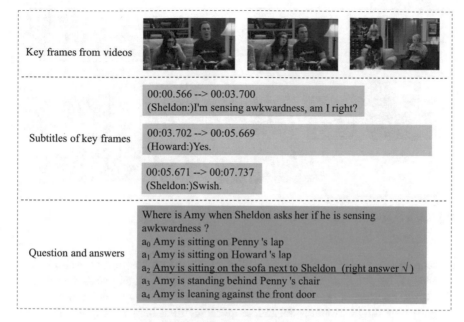

Fig. 8.1 Sample entry from the TVQA dataset

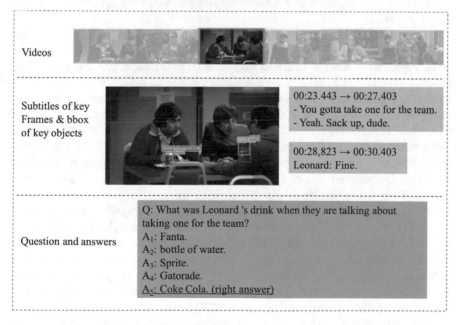

Fig. 8.2 Sample entry from the TVQA+ dataset

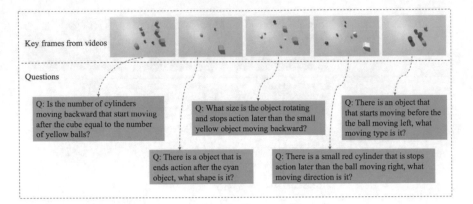

Fig. 8.3 Sample entry from the SVQA dataset

accompanied by a JSON file recording the attributes and position of each involved geometry. The questions in SVQA are generated by predefined question templates. The key properties of questions are the exclusively long length and compositional property regarding various spatial and temporal relations between objects. The questions in the SVQA can be decomposed into a logical tree or chain layouts, in which each node can be regarded as a subtask requiring a reasoning operation, i.e., filter shape (Fig. 8.3).

CLEVRER [13] is a diagnostic video dataset to perform the systematic evaluation of computational models on a wide range of reasoning tasks. CLEVRER includes 20,000 synthetic videos of colliding objects and more than 300,000 questions and answers. Videos are generated by a physics engine, including three shapes, two materials, eight colors and three types of events: enter, exit and collision. The dataset offers object attributions, event annotations and object motion traces as annotations. CLEVRER includes four types of questions: descriptive (e.g., "what color"), explanatory ("what responsible for"), predictive ("what will happen next") and counterfactual ("what if"), with 219,918 descriptive questions, 33,811 explanatory questions, 14,298 predictive questions and 37,253 counterfactual questions. Each question is represented by a tree-structured functional program.

AGQA [1] is a benchmark to assess the compositional spatiotemporal reasoning ability. The dataset contains a balanced 3.9M and an unbalanced 192M question/answer pair associated with 9.6K videos. The video source is from Charades, and annotations are from Charades?action annotations and Action Genome spatiotemporal scene graphs, which ground all objects with bounding boxes and actions with timestamps in the video. Questions are generated by handcrafted programs that operate over these annotations. Moreover, AGQA also provides three new compositional spatiotemporal splits—(i) novel compositions, (ii) indirect references and (iii) additional compositional steps—for testing the reasoning ability of the model. AGQA is a highly challenging benchmark as it is built upon real video data sources and generated from complex question templates.

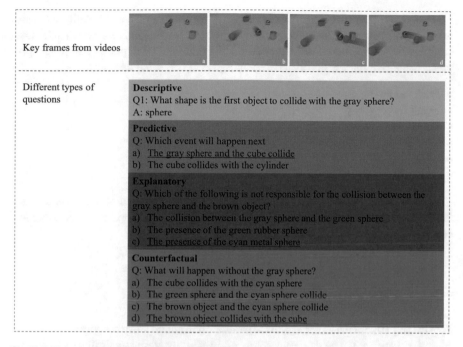

Fig. 8.4 Sample entry from the CLEVRER dataset

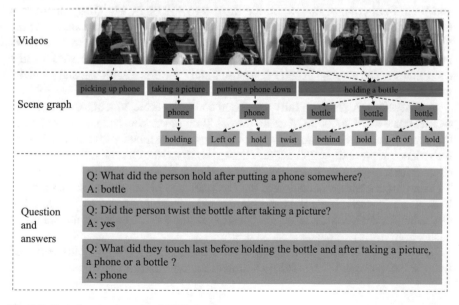

Fig. 8.5 Sample entry from the AGQA dataset

Traffic QA [11] is a diagnostic benchmark for the cognitive capability of causal inference and event understanding models in complex traffic scenarios. The dataset contains 10,080 in-the-wild videos and 62,535 annotated QA pairs. Traffic QA proposes 6 challenging traffic-related reasoning tasks (Figs. 8.4 and 8.5):

- Basic understanding
- Event forecasting
- Reverse reasoning
- Counterfactual inference
- Introspection
- Attribution

QA pairs are designed by annotators related to the 6 tasks above. The average length of questions is 8.6 words. Various levels of spatiotemporal understanding and causal reasoning are required for this dataset.

8.2.2 Single-Step Reasoning Dataset

Movie QA [10] is a dataset that aims to evaluate automatic story comprehension from both video and text. The dataset consists of 14,944 questions regarding 408 movies. A key property of the dataset is that it contains video clips, plots, subtitles, scripts and DVS. Furthermore, for 140 of 408 movies (6,462 of 14,944 QAs), the dataset has timestamp annotations indicating the location of the question and answer in the video. The multichoice questions contain who? did what? to whom? why? and how? certain events occurred, and the multichoice questions have 4 wrong answers and 1 right answer. The average length of questions and answers is approximately 9 and 5 words, respectively.

ActivityNet-QA [14] is a fully annotated and large-scale VideoQA dataset. The dataset consists of 58,000 QA pairs on 5,800 complex web videos derived from the popular ActivityNet dataset, which contains approximately 20,000 untrimmed web videos representing 200 action classes. ActivityNet-QA includes three types of questions, pertaining to motion, spatial relationship and temporal relationship. To avoid improper representation of the QA pairs, the maximum question length is limited to 20 words, while the maximum answer length is limited to 5 words. The QA pairs are designed by separated question annotators and answer annotators to ensure high quality of the dataset.

TGIF-QA [2] is a large-scale dataset consisting of 103,919 QA pairs collected from 56,720 animated GIFs. TGIF-QA involves four types of tasks:

Videos				
Question	Q:How many times does the cat lick?	Q: What does the duck do 3 times?	Q: What does the bear on right do after sitting?	Q: What is dancing in the cup?

Fig. 8.6 Sample entry from the TGIF-QA dataset

1. Repetition count, with 11 possible answers from 0 to 10+.
2. Repeating action, which is in a multiple-choice question format. Each question has 5 potential answers.
3. State transition, which queries the transitions of certain states.
4. Frame QA, which can be answered from one of the frames in a video.

Questions are generated automatically based on several manually designed templates (Fig. 8.6).

MarioQA [7] is a synthetic dataset of events that uses Super Mario gameplay videos with their logs. Each entry in the dataset consists of a 240?20 video clip containing multiple events and a question with the answer. A total of 187,757 examples are collected from 13 h of gameplay. There exist 92,874 unique QA pairs, and each video clip contains 11.3 events on average. The question/answer pairs are generated based on 11 distinct events: kill, die, jump, hit, break, appear, shoot, throw, kick, hold and eat. The generated questions are categorized into three types: event-centric, counting and state questions. The dataset contains three subsets with different temporal relationships:

1. Questions with no temporal relationship (NT).
2. Questions with easy temporal relationships (ET).
3. Questions with hard temporal relationships (HT).

In particular, NT, ET and HT pertain to queries regarding unique events in the entire video with temporal relationships, globally unique events, and distracting events, respectively. Notably, the dataset is designed for reasoning the temporal dependency and understanding the temporal relationships between video events.

PororoQA [3] is a dataset built on the video series for a children cartoon. The dataset contains 16,066 scene-dialogue pairs of 20.5 h videos, 27,328 fine-grained sentences for scene description and 8,913 story-related QA pairs. Since the video source is a children's cartoon, the dataset has a simpler background and clearer events than those built on movies and TVs, which facilitates the understanding of the video. The video series contains 171 episodes, and the average length of the videos is 7.2 min. The description sentences and question/answer pairs have been manually collected by annotators from the Amazon Mechanical Turk (AMT) platform. The dataset contains 11 types of questions:

1. Action
2. Person
3. Abstract
4. Detail
5. Method
6. Reason
7. Location
8. Statement
9. Causality
10. Yes/No
11. Time

The average length of the scene description is 13.6 words.

8.3 Traditional Video Spatiotemporal Reasoning Using an Encoder-Decoder Framework

The core of video question answering is video spatiotemporal reasoning. The fundamental pipeline can be described as follows. First, video is represented as features of different levels, including object-level, frame-level and clip-level features. For object-level feature extraction, most researchers adopt Faster R-CNN to detect local parts of videos. Frame-level features are coarse-grained representations of global visual information that capture more information than object-level features, such as scenes, and ResNet and VGGNet are widely used to extract such features. Clip-level features capture the information conveyed by several frames (e.g., actions), and C3D networks are frequently used to extract such features. Second, text is represented as features of different levels, including sentence- and word-level features. Word embedding techniques such as word2vec and GloVe are widely used for extracting word-level features, while skip-thought and BERT are used to extract sentence-level features. After obtaining the visual and text features, the model implements video spatiotemporal reasoning on the input features, thereby obtaining a contextual representation. Finally, the contextual representation is input to the answer generation unit, which is usually a discriminative or multiclass classification model.

Since videos and questions are naturally in a sequential format, the encoder-decoder model, which is widely used in machine translation applications, can effectively realize video spatiotemporal reasoning.

Zhu et al. [17] used a GRU to learn the temporal structures of videos and designed a dual-channel ranking loss to answer multiple-choice questions. As shown in Fig. 8.7, the authors first trained three encoder-decoder models to learn the past, present and future representation of the input frames and separately trained the three models. In this context, the main purpose of the encoder-decoder model is to reconstruct the input frame to ensure that the encoder can better represent the frames. The structure of the decoder is similar to that of the encoder. Subsequently, a dual-channel ranking

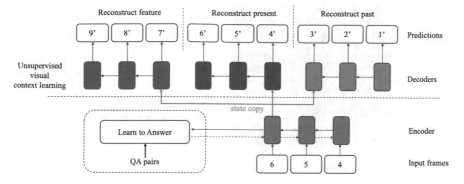

Fig. 8.7 Encoder-decoder model (top). Learning to answer questions (bottom)

loss is adopted to calculate the similarity between the visual contextual representation and candidate choice of the question, which can be expressed as follows:

$$Loss = \min_{\theta} \sum_{v} \sum_{j \in K, j \neq j'} \lambda \ell_{word} + (1 - \lambda) \ell_{sent}, \lambda \in [0, 1]$$

$$\ell_{word} = \max \left(0, \alpha - v_p^T p_{j'} + v_p^T p_j\right)$$

$$\ell_{sent} = \max \left(0, \beta - v_s^T s_{j'} + v_s^T s_j\right),$$

where $v_p = W_{vp}v$, $v_s = W_{vs}v$ and $p_j = W_{pv}y_j$, $s_j = W_{sv}z_j$. Here, v is the visual representation learned from the encoder-decoder model, and y and z are the textual representations. The final answer is the candidate with the highest similarity. Other researchers [17] used only the GRU to realize the temporal reasoning of videos. In this framework, the model can capture the information of the video in a long duration. However, the answer generation and video representation are separately trained, and thus the ability to reason the relation between the text and video is inferior.

Although the abovementioned studies used the basic encoder-decoder framework, reasoning between textual and visual information could not be realized. Certain researchers added simple attention to the model either in the encoder or decoder to examine the relations between different modalities of information.

Lei et al. [4] proposed a multistream end-to-end trainable neural network. This model takes different contextual sources, including regional visual features, visual concept features and subtitles, along with question/answer pairs as inputs to each stream. The video is represented by three features:

1. Regional visual features, which are top-K regions detected by Faster R-CNN in each frame.
2. Visual concept features, which are detected labels including both objects and attributes.
3. ImageNet features, which are extracted by ResNet101.

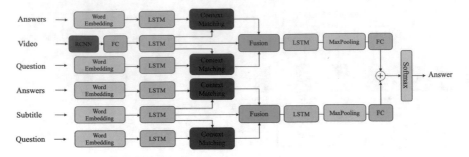

Fig. 8.8 Multimodal model

All the sequential information, including text and visual information, is encoded using a bidirectional LSTM, in which the hidden states are concatenated to serve as visual and textual representations. Subsequently, a context matching module, which is a context-query attention layer, is adopted to generate video-guided-question and video-guided-answer representations, which are fused as the contextual input of the answer generation layer. In another approach [4], multimodal information is fully used, as shown in Fig. 8.8, which enriches the contextual representation. Furthermore, the context matching unit yields richer relations between the textual and visual information.

Jang et al. [2] proposed a dual-LSTM-based model with spatial and temporal attention frameworks. As shown in Fig. 8.9, first, both frame-level and sequence-level video features are extracted by ResNet pretrained on the ImageNet 2012 classification dataset and C3D pretrained on the Sport1M dataset, which are concatenated as the visual representation. Questions and answers are embedded as two sequences. Three dual LSTMs are applied as separate encoders for the visual, question and answer representations. Before the visual representations are input to the LSTM, an attention unit is adopted, as shown on the left side of Fig. 8.10, which considers the visual features with encoded text representations to determine which regions in a frame are most related to the questions and answers. Furthermore, another attention unit, as shown on the right of Fig. 8.10, is used to learn the frames that must be examined in a video, thereby considering the sequential visual hidden states from dual-LSTM with encoded textual representations. Since the dataset used to train the model has three types of answers (multichoice, open-ended number and open-ended word), the proposed model especially trains three decoders to generate answers along with two attention units to implement spatiotemporal reasoning over the video information associated with the question/answer pairs.

Xue et al. [12] proposed a set of models, as shown in Fig. 8.11, including the following three models:

1. A sequential video attention model, as shown in the top left part of Fig. 8.11.
2. A temporal question attention model, as shown in the top right part of Fig. 8.11.
3. A decoder for answer generation, as shown in the bottom part of Fig. 8.11.

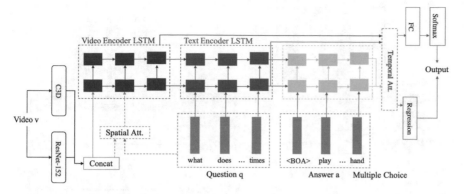

Fig. 8.9 ST-VQA model

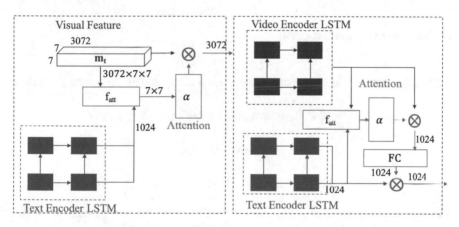

Fig. 8.10 Spatial attention (left). Temporal attention (right)

The sequential video attention model and temporal question attention model are in a dual format. The sequential video attention model considers the video representations encoded by LSTM with a sequence of question hidden states from LSTM, and the final accumulated representation is the visual encoding of this model. The output $V = r(T)$ can be expressed as follows:

$$r(i) = y_v^T s_v(i) + \tanh \left(V_{rr} r(i-1) \right), \quad 1 \leq i \leq T$$

$$s(i, j)_v \propto \exp \left(W_{cs}^T c(i, j) \right)$$

$$c(i, j) = \tanh \left(W_{vc} y_v(j) + U_{qc} y_q(i) + V_{rc} r(i-1) \right),$$

where $y_v(j)$ is the j_{th} frame feature, and $y_q(i)$ is the i_{th} text feature. The temporal question attention model considers the question representations encoded by LSTM with a sequence of video hidden states from the LSTM, and the final representation is

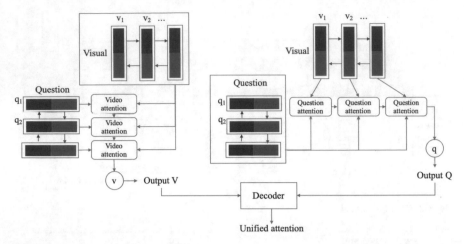

Fig. 8.11 Model for unified attention

the textual encoding of this model. The output $q = w_T$ can be expressed as follows:

$$w(j) = y_q^T s(j)_t + \tanh\left(V_{ww}w(j-1)\right), \quad 1 \le j \le N$$

$$s(j,i)_t \propto \exp\left(U_{cs}^T c(j,i)\right)$$

$$c(j,i) = \tanh\left(W_{qc}y_q(i) + U_{vc}y_v(j) + V_{wc}w(j-1)\right).$$

Subsequently, the two types of encodings are fused and fed into the decoder, which is a two-layer LSTM, to generate the open-ended answer sequence.

Zhao et al. [15] proposed a hierarchical spatiotemporal attentional encoder-decoder learning method with a multistep reasoning process to realize open-ended video question answering. As shown in Fig. 8.12, first, a multistep spatiotemporal attentional encoder network is developed to learn the contextual representation of the video and question. Similar to the introduced models, in each step, the model first uses the spatial attention model to localize the targeted regions in each frame attended with the question. For the i_{th} object in the j_{th} frame, the spatial attention score $s_{ji}^{(s)}$ is defined as

$$s_{ji}^{(s)} = \mathbf{w}^{(s)} \tanh\left(\mathbf{W}_{qs}\mathbf{q} + \mathbf{W}_{fs}\mathbf{f}_{ji} + \mathbf{b}_s\right).$$

Subsequently, the spatially attended frame representation is expressed as

$$\mathbf{v}_j^{(s)} = \sum_i \alpha_{ji}\mathbf{f}_{ji}$$

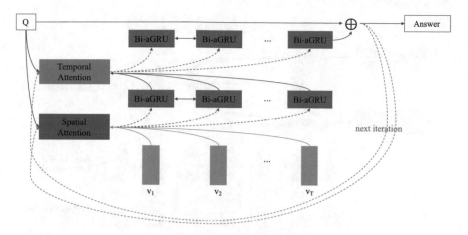

Fig. 8.12 Overview of open-ended video question answering via a hierarchical spatiotemporal attentional encoder-decoder learning framework

$$\alpha_{ji} = \frac{\exp\left(s_{ji}^{(s)}\right)}{\sum_i \exp\left(s_{ji}^{(s)}\right)}.$$

In the presence of redundant and multiple frames, it is important to localize the relevant video frames, and the temporal attention model is operated over the videos, thereby localizing the target frames in the video. The relevance scores of the s_{th} hidden state are expressed as

$$s_j^{(t)} = \mathbf{w}^{(t)} \tanh\left(\mathbf{W}_{qt}\mathbf{q} + \mathbf{W}_{ht}\mathbf{h}_j^{(s)} + \mathbf{b}_t\right).$$

Subsequently, the aGRU updates the current hidden state by

$$\mathbf{h}_j^{(t)} = \beta_j \odot \tilde{\mathbf{h}}_j^{(t)} + \left(1 - \beta_j\right) \odot \mathbf{h}_{j-1}^{(t)}$$

$$\beta_j = \frac{\exp\left(s_j^{(t)}\right)}{\sum_j \exp\left(s_j^{(t)}\right)}.$$

The information of the relevant video frames is embedded into the hidden states according to the abovementioned process. To learn better visual and textual representations, the representations are recursively updated following the updating formula.

Zhao et al. [16] proposed an adaptive hierarchical reinforced encoder-decoder network to address a long-form video question answering task. The adaptive recurrent neural networks with the binary gate function segment the frame-level features extracted using ConvNet and decide whether the hidden state and memory cell at

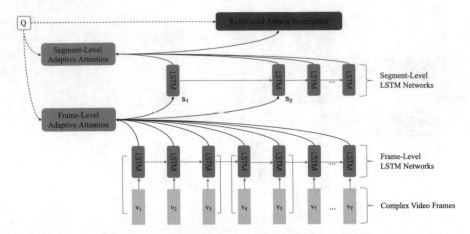

Fig. 8.13 Framework of adaptive hierarchical reinforced networks for open-ended long-form video question answering

timestamp t must be transferred to the next timestamp $t + 1$. The binary gate function, which segments the frame features during the encoding process, calculates γ_t to identify the similarity between the hidden state at timestamp t and visual representation at timestamp $t + 1$. Given the semantic representation $\{h_1, h_2, \ldots, h_N\}$ with binary gate values $\{\gamma_1, \gamma_2, \ldots, \gamma_N\}$, joint question-attended video segment representations are learned, which are then input to segment-level LSTM networks to generate semantic representations, denoted as $\{h_1^s, h_2^s, \ldots, h_K^s\}$. The decoder is designed as a reinforced neural network, and it generates open-ended answers based on the similarity between the semantic and question representations. The key contributions include the development of an adaptive hierarchical encoder to learn segment-level question-aware video representations and the formulation of a reinforced decoder to generate answers (Fig. 8.13).

As discussed previously, traditional video spatiotemporal reasoning uses an encoder-decoder framework as their basic framework. Although other techniques, e.g., attention, have been applied, detailed attention structures cannot be designed and deeper relations associated with the multimodal information cannot be explored.

References

1. M. Grunde-McLaughlin, R. Krishna, M. Agrawala, AGQA: a benchmark for compositional spatio-temporal reasoning, in *Proceedings of the IEEE/CVF Conference on Computer Vision and Pattern Recognition* (2021), pp. 11287–11297
2. Y. Jang, Y. Song, Y. Yu, Y. Kim, G. Kim, TGIF-QA: toward spatio-temporal reasoning in visual question answering, in *Proceedings of the IEEE Conference on Computer Vision and Pattern Recognition* (2017), pp. 2758–2766

3. K.-M. Kim, M.-O. Heo, S.-H. Choi, B.-T. Zhang, DeepStory: video story QA by deep embedded memory networks (2017), arXiv:1707.00836
4. J. Lei, L. Yu, M. Bansal, T.L. Berg, TVQA: localized, compositional video question answering (2018), arXiv:1809.01696
5. J. Lei, L. Yu, T.L. Berg, M. Bansal, TVQA+: spatio-temporal grounding for video question answering (2019), arXiv:1904.11574
6. M. Malinowski, M. Fritz, A multi-world approach to question answering about real-world scenes based on uncertain input. Adv. Neural Inf. Process. Syst. **27**, 1682–1690 (2014)
7. J. Mun, P. Hongsuck Seo, I. Jung, B. Han, MarioQA: answering questions by watching game-play videos, in *Proceedings of the IEEE International Conference on Computer Vision* (2017), pp. 2867–2875
8. X. Song, Y. Shi, X. Chen, Y. Han, Explore multi-step reasoning in video question answering, in *Proceedings of the 26th ACM International Conference on Multimedia* (2018), pp. 239–247
9. G. Sun, L. Liang, T. Li, B. Yu, M. Wu, B. Zhang, Video question answering: a survey of models and datasets. Mob. Netw. Appl. 1–34 (2021)
10. B. Wang, Y. Xu, Y. Han, R. Hong, Movie question answering: remembering the textual cues for layered visual contents, in *Thirty-Second AAAI Conference on Artificial Intelligence* (2018)
11. L. Xu, H. Huang, J. Liu, SUTD-TrafficQA: a question answering benchmark and an efficient network for video reasoning over traffic events, in *Proceedings of the IEEE/CVF Conference on Computer Vision and Pattern Recognition* (2021), pp. 9878–9888
12. H. Xue, Z. Zhao, D. Cai, Unifying the video and question attentions for open-ended video question answering. IEEE Trans. Image Process. **26**(12), 5656–5666 (2017)
13. K. Yi, C. Gan, Y. Li, P. Kohli, J. Wu, A. Torralba, J.B. Tenenbaum, CLEVRER: collision events for video representation and reasoning (2019), arXiv:1910.01442
14. Z. Yu, D. Xu, J. Yu, T. Yu, Z. Zhao, Y. Zhuang, D. Tao, ActivityNet-QA: a dataset for under-standing complex web videos via question answering, in *Proceedings of the AAAI Conference on Artificial Intelligence*, vol. 33 (2019), pp. 9127–9134
15. Z. Zhao, Q. Yang, D. Cai, X. He, Y. Zhuang, Video question answering via hierarchical spatio-temporal attention networks, in *IJCAI* (2017), pp. 3518–3524
16. Z. Zhao, Z. Zhang, S. Xiao, Z. Yu, J. Yu, D. Cai, F. Wu, Y. Zhuang, Open-ended long-form video question answering via adaptive hierarchical reinforced networks, in *IJCAI*, vol. 3 (2018), p. 4
17. L. Zhu, Z. Xu, Y. Yang, A.G. Hauptmann, Uncovering the temporal context for video question answering. Int. J. Comput. Vis. **124**(3), 409–421 (2017)

Chapter 9
Advanced Models for Video Question Answering

Abstract In Chap. 8, we introduced several traditional models for video question answering tasks based on the encoder-decoder framework. However, other models exist beyond this framework, which exhibit fine architectures and performances. In this chapter, we categorize these methods into four categories, i.e., *attention on spatiotemporal features, memory networks, spatiotemporal graph neural networks* and *multitask pretraining* and discuss the characteristics of these frameworks.

9.1 Attention on Spatiotemporal Features

As described in Chap. 2, the attention model can be described as a function that maps a query and a set of key-value pairs to an output [16], where the query, keys, values and output are vectors. In the video question answering task, the visual and textual representations usually serve as queries, keys and values to obtain question-attended video representations and video-attended question representations at different levels. As described in Sect. 8.3, the models usually apply simple spatiotemporal attention over feature vectors. In this section, we introduce models that adopt a more complex attention architecture to employ deeper reasoning.

Xu et al. [17] proposed an end-to-end model that gradually refines its attention over the appearance and motion features of the video attended with questions. First, feature extraction is performed, in which frame-level features are extracted by the VGG network, clip-level features are extracted by the C3D network and word-level features are extracted. The proposed model designs an attention unit named the attention memory unit (AMU), which contains the following four subunits:

1. Attention (ATT), which attends visual features with contextual representations.
2. Channel fusion (CF), which scores the frame-level feature representation and clip-level feature representation and fuses them according to the score.
3. Memory (LSTM), which uses the sum of (i) the fused representation, (ii) hidden state h_t of the word encoder and iii) visual representation v_{t-1} produced by AMU at timestamp t-1, as the input of an LSTM, thereby controlling the input of the second attention operation and remembering the attention history.

Q. Wu et al., *Visual Question Answering*, Advances in Computer Vision and Pattern Recognition, https://doi.org/10.1007/978-981-19-0964-1_9

4. Refine (REF), which performs the final attention over the features to obtain the final visual representation v_t at timestamp t.

Each question is processed per word at each timestamp, and the word embedding and hidden states of the word encoder are input to the AMU unit until all the words have been processed to generate the final refined contextual representation. Finally, contextual representation v_T, question memory vector c_T^q and AMU memory vector c_T^a are input to a softmax classifier to generate an answer. The contribution of Xu et al.'s research [17] is the design of an attention unit (AMU) that can gradually refine the attention over information, thereby identifying richer spatiotemporal relations.

The abovementioned models treat the words in the questions in a similar manner and cannot differentiate the words before they are input to the models. Xue et al. [18] proposed the heterogeneous tree-structured memory network (HTreeMN), which is based upon the syntax parse trees of the question sentences. HTreeMN defines two types of words as follows:

1. Visual words, which are combined with visual features.
2. Verbal words, in which the former words are processed with an attention module while the latter words are not.

First, a parse tree is developed according to the grammatical structure of the question by using the *StanfordParser* tool, in which nodes with visual words as children are regarded as visual nodes and other words are treated as verbal nodes. In the visual node, temporal attention is produced over the video frames, and the output is passed to the father node. In contrast, the verbal node implements a linear operation. The whole network is processed in a bottom-up manner to generate a contextual representation. Finally, the softmax function in the root node helps generate the answer.

Although the abovementioned models use single-hop reasoning, certain approaches employ multihop reasoning to extract richer contextual representations. Mun et al. [13] proposed a neural network containing three components: question embedding, video embedding and classification networks. The question embedding network uses a pretrained question embedding network to process the sequential information, and the video embedding network uses a 3D fully convolutional network to generate visual features. In addition to the traditional question attending single-step temporal attention, the approach proposed by Mun et al. [13] utilizes a multistep temporal attention mechanism, in which the question attending embedding at timestamp t is refined by adding the previously attended embedding. Furthermore, spatiotemporal attention over frame features and fully connected layers is employed to obtain two more contextual representations. Finally, a classification network is designed to use the fused representation of the three representations to generate an answer.

In contrast to the simple application of multihop reasoning on temporal attention, certain researchers employ multihop reasoning over a more complex unit in the model. Song et al. [15] proposed a refined GRU known as temporal attention-GRU (ta-GRU) to capture the long-term temporal dependency and gather complete visual cues. ta-GRU is a refined GRU whose hidden state transfer processes attention over the entire historical hidden state attended with the question to strengthen the long-term temporal dependency.

Le et al. [10] proposed an end-to-end layered architecture, which is composed of a question-guided video representation layer and a generic reasoning layer to produce answers. The model contains the following three components:

1. Hierarchical video representation with CRN.
2. Visual multistep reasoning with MAC cells.
3. Answer decoders.

First, a video is represented as clip-level features, with each clip consisting of several frame features. Later, question-attended clip-level features are placed in a clip-based relation network (CRN) to generate the relation between different clip segments, the sizes of which are 2, 3 and 4 clips. Subsequently, memory-attention-composition (MAC) networks are used to implement multistep reasoning on the video contextual representation. Finally, regression or classification is employed according to the question type to generate an answer. The main contribution of Le et al.'s work [10] is a hierarchical model that can identify spatiotemporal relations at the clip level.

Le et al. [11] designed a novel neural structure to identify nonadjacent relations, named the conditional relation network (CRN), which can universally serve as neural units in the model. The reusable computation unit of the CRN takes an array of objects S (e.g., frames and clips) and conditioning feature c as inputs and outputs the k-tuple conditional relations of the input, where $k = 1, 2, ..n - 1$, and n is the size of the input array. Subsequently [11], a hierarchical conditional relation network (HCRN) is built upon a CRN unit, which consists of a clip-level processing network and video-level processing network. For the clip-level processing network, each clip is processed by a dual layer CRN unit conditioned by clip-level motion and questions separately, while all the outputs from the clip-level network form an input sequence of the video-level network, processed by a dual layer CRN unit separately conditioned by video-level motion and questions. Finally, the contextual representation of the network is input to a softmax function to generate an answer.

The abovementioned approaches adopt RNNs to encode sequential information. Notably, these models are time-consuming and cannot easily model long-range dependencies. Therefore, certain researchers use self-attention instead. Li et al. [12] proposed a new structure, named positional self-attention with coattention (PSAC), which does not require RNNs for video question answering. The architecture consists of three key components: video-based positional self-attention block (VPSA), question-based positional self-attention block (QPSA) and video question-coattention block (VQ-Co), among which the former two blocks share the same positional self-attention structure. The positional self-attention model calculates the weight distribution at each position by attention over all positions within the same sequence and adds the representation of the absolute positions. First, the model calculates the positional self-attended frame features and self-attended textual features. Subsequently, the visual-attended textual representation and textual-attended visual representation are concatenated as the final contextual representation. Finally, the contextual representation of the network is input to a softmax function to generate an answer.

Many models autonomously learn all the knowledge from the supervised information, while other models refer to an external knowledge base according to the content of the video and question. Jin et al. [7] developed a question knowledge-guided progressive spatiotemporal attention network to learn the joint video representation for video question answering tasks. Because the model refers to the external knowledge base, it can grasp the knowledge that cannot be learned during the training process (e.g., common sense), which strengthens the ability of the models.

9.2 Memory Networks

RNNs and their variants are usually used for memorizing sequential information; however, this framework may not be adequate for long-term information (e.g., video). As discussed in Chap. 2, a memory network that is equipped with a more powerful memory has been widely used to solve the VideoQA task. In the following section, we describe models based on the memory network and its variants, such as the end-to-end memory network and dynamic network.

Kim et al. [9] developed a video-story learning model known as deep embedded memory network (DEMN) to reconstruct stories from a joint scene-dialog video stream by using a latent embedding space of observed data. For a given question, an LSTM-based attention model uses the long-term memory to recall the best question-story-answer triplet by focusing on specific words containing key information. DEMN takes clip-level features and textual features describing the clip as inputs and modifies the generalization component of the memory networks to generate the story description, and these entities are stored in the memory array in order. Therefore, the memory array represents the entire video content. The output feature map component utilizes the question-guided attention model to generate the question-attended video story. Finally, this story concatenated with the question is the input of the response component to generate the answer.

In contrast to the use of a memory array as a sequence, Fan et al. [1] employed an attention model over the memory array. The authors propose a heterogeneous memory network that processes heterogeneous visual features such as motion and appearance visual features, thereby generating global context-aware visual and textual features, respectively, by interacting current inputs with memory contents. The heterogeneous visual memory unit contains the following three main parts:

1. Memory array.
2. Multiple read heads and write heads enable multiple inputs and outputs, including encoded motion features and appearance features.
3. Three hidden states, which store the motion contents, appearance contents and global context-aware features.

The write operation calculates the weight of the appearance, motion and memory array based on hidden states and updates the contents of memory arrays according to the weight values associated with the inputs. The read operation determines the weight for each memory unit according to the hidden state and content of appearance and motion. Subsequently, the model fuses the contents from the memory array according to the weights to generate a contextual representation. Next, the hidden states are updated according to the current hidden states and contextual representation. The question is processed in the same manner by another external question memory with a single write head, read head and hidden state. Subsequently, the two contextual representations are fused and concatenated with a question-attended video representation to generate the answer using the softmax function.

Na et al. [14] proposed a novel memory network model named the ReadWrite Memory Network (RWMN), which is trained to store the movie content with proper representation in the memory, extract relevant information from memory cells in response to a given query and select correct answers from five choices. The *write* network employs a convolutional neural network (CNN) to jointly learn adjacent embedding into a memory array. In the read network, the memory array is first transformed into a query-dependent memory by CBP upon query embedding, and another CNN is used to reconstruct a series of scenes as a whole. Finally, the answer-generated unit uses both query embedding and memory reconstructed by the read network to generate answers.

As discussed in Chap. 2, MemNN-based models cannot be trained in an end-to-end manner. Therefore, the RWMN [14] uses the MemN2N model, thereby refining the input of the model to obtain the extended end-to-end memory networks (E-MN), which can capture the temporal relation among actions in the succeeding frames. The Bi-LSTM is used to encode the input sequence, and the encoded sequences are input to the MemN2N model.

Gao et al. [2] developed a model based on the architecture of a dynamic memory network (DMN). The model, named motion-appearance comemory networks, modifies the input module to generate a contextual representation and changes the episodic memory module into a motion-appearance comemory module. The input module considers multiple streams of features as the input and uses the temporal convolutional layers to model the temporal contextual information and deconvolutional layers to recover temporal resolution, therefore, building multiple levels of temporal representations $F_L = F_L^1, F_L^2, ..., F_L^N$. To process both appearance and motion representations, the motion-appearance comemory module contains two separate memory modules: m_b^t for motion memory and m_a^t for appearance memory at iteration t. At iteration t, the motion-appearance comemory module consists of three steps: comemory attention, dynamic fact ensemble, and memory update. The comemory attention motion memory uses the appearance memory and question representation to separately assign attention over motion facts and appearance facts. The dynamic fact ensemble calculates the weighted facts based on the attention gate calculated by comemory attention. Subsequently, the motion memory and appearance memory are separately updated according to the ensemble fact, current memory and question.

Finally, the answer module uses a linear regression function that takes the memory state and outputs the scores of answer candidates.

Kim et al. [8] proposed a progressive attention memory network (PAMN), which contains the following four main components:

1. Memory embedding module.
2. Progressive attention mechanism, which pinpoints temporal parts that are relevant to answering the question.
3. Dynamic modality fusion.
4. Answer generation.

The memory embedding module uses a feed-forward neural network (FFN) to separately generate video and subtitle embedding into memory. Subsequently, the progressive attention mechanism takes dual memory, question embedding and answer embedding as inputs and progressively attends and updates the dual memory. The dynamic modality fusion module attends the memory with question embedding separately, generating information from the necessary modality. Finally, a belief correction answering scheme considers the contextual memory and generates answers.

9.3   Spatiotemporal Graph Neural Networks

In addition to the abovementioned deep learning-based models, graph-based models exist. Although the number of such models is considerably smaller than those involving deep structures, graph-based models can achieve promising results. Graphs are usually built upon videos [3, 4, 6], and graphs generated from both videos and questions also exist [5]. The most challenging task for graph-based models is to extract the attributed graph(s) based on rich information and relations inside the videos and questions.

Gu et al. [3] proposed a graph-based relation-aware neural network to explore a more fine-grained visual representation, which could explore the relationships and dependencies between objects spatially and temporally. First, a graph-based video representation is generated, including objects and object relations. All objects from all frames in the video are defined as nodes $N = \{n_i\}$ of the graph, while edges are formed by two kinds of edges, including iteration edges E_R, which (i) link all object pairs in the same frames and trajectory edges, and (ii) link nodes of the same object that have different positions and appearances over time calculated by the identification score [19]. The initial states of nodes and edges are defined as object appearance features and spatial relations between two connected objects. Subsequently, the hidden state of the graph is updated considering interactive edges and trajectory edges, and the final representation of the graph is a list of the hidden states of all the nodes. In addition, the authors propose a multihead attention mechanism named multi-interaction to capture both elementwise and segmentwise sequence interactions to generate question-attended video representations, question representations, and question-attended graphical representations. Finally, the question-attended video representations and question-attended graphical representations are input to the answer module to generate answers.

In contrast to this model, Huang et al. [4] used graph convolutional networks (GCNs) to develop a model named location-aware graph convolutional networks, which explores location and relations among object interactions. A location-aware graph is a fully connected graph built on K detected objects for each frame as nodes. The hidden state of each node is combined with both appearance and location features. P-layer graph convolutions are operated on the location-aware graph, the p-th layer of which can be represented as $X^{(p)} = A^{(p)} X^{(p-1)} W^{(p)}$, where $X^{(p)}$ is the hidden state of the p-th layer, $A^{(p)}$ is the adjacency matrix calculated from the node features in the p-th layer and $W^{(p)}$ is the trainable weight matrix, outputting the regional features F^R calculated by $F^R = X^{(p)} + X^{(0)}$. Subsequently, visual features are generated concatenating regional and appearance features. Visual features, together with textual features extracted from questions, are used to generate answers.

Jin et al. [6] proposed an adaptive spatiotemporal graph-enhanced vision-language representation learning model to address the VideoQA task, which consists of the following two components:

- Adaptive spatiotemporal graph module for dynamic object representation learning.
- Vision-language transformer module for multimodal representation.

The adaptive spatiotemporal graph module generates the spatiotemporal representations in two steps. First, a spatial graph is built for each frame, the nodes of which are detected objects. The hidden states of the nodes are generated by appearance features, object class information and geometric information, and the adjacent matrix is the affinity of hidden states of nodes. A GCN is applied on each frame-based spatial graph, generating updated embedding of object nodes $V = \{v_i\}$ of object nodes. Second, a spatiotemporal graph is progressively built. Starting from the object nodes in the spatial graph of the first frames, denoted as anchor nodes, an anchor tube set $A = \{a_1, a_2, ..a_n | a_i = \{v_i\}\}$ is initialized, which observes each object node v_j in each subsequent frame to update the anchor tube set A according to the similarity score between v_j and the anchor node. If the object node v_j exhibits a high similarity with the anchor node v_i, v_j is added to a_i; otherwise, v_j is marked as a new anchor tube by adding $a_j = \{v_j\}$ to the anchor tube set A. After all the frames are considered, anchor tubes from the nodes of the spatiotemporal graph and adjacent matrix represent the similarity matrix of the graph. Another GCN layer is employed on this graph to capture temporal dynamic information of each spatiotemporal tube, and the final representation is a set of hidden states for all the nodes. Next, a vision-language transformer considers final visual representations from the spatiotemporal graph and question embedding as the input to generate contextual representations for generating answers.

The abovementioned approaches generate graphs using only the information from videos. In contrast, the approach proposed by Jiang et al. [5] fuses information from both questions and videos to build a multimodal graph. The authors propose a deep heterogeneous graph alignment network over video shots and question words. A coattention transformation is implemented on the visual features and textual features encoded by the GRU to obtain question-attended visual features and video-attended textual features, in which the feature dimensions are the same and concatenated as a heterogeneous input matrix X. The heterogeneous graph uses vectors in the matrix $X = \{x_i\}$ as nodes, and the adjacency matrix G is calculated as $G = \phi(X)\phi(X)^T$, where ϕ is a learnable transformation for alignment. Later, a one-layer graph convolutional network is operated on the heterogeneous graph, followed by self-attention pooling to obtain a local result vector that reflects the underlying cross-modal relations after the local reasoning procedure. The local reasoning result, together with the global reasoning result calculated by a bilinear fusion module over the last hidden states of the visual and linguistic GRU encoder, is used to generate the answer.

References

1. C. Fan, X. Zhang, S. Zhang, W. Wang, C. Zhang, H. Huang, Heterogeneous memory enhanced multimodal attention model for video question answering, in *Proceedings of the IEEE/CVF Conference on Computer Vision and Pattern Recognition* (2019), pp. 1999–2007
2. J. Gao, R. Ge, K. Chen, R. Nevatia, Motion-appearance co-memory networks for video question answering, in *Proceedings of the IEEE Conference on Computer Vision and Pattern Recognition* (2018), pp. 6576–6585
3. M. Gu, Z. Zhao, W. Jin, R. Hong, F. Wu, Graph-based multi-interaction network for video question answering. IEEE Trans. Image Process. **30**, 2758–2770 (2021)
4. D. Huang, P. Chen, R. Zeng, Q. Du, M. Tan, C. Gan, Location-aware graph convolutional networks for video question answering, in *Proceedings of the AAAI Conference on Artificial Intelligence*, vol. 34 (2020), pp. 11021–11028
5. P. Jiang, Y. Han, Reasoning with heterogeneous graph alignment for video question answering, in *Proceedings of the AAAI Conference on Artificial Intelligence*, vol. 34 (2020), pp. 11109–11116
6. W. Jin, Z. Zhao, X. Cao, J. Zhu, X. He, Y. Zhuang, Adaptive spatio-temporal graph enhanced vision-language representation for video QA. IEEE Trans. Image Process. (2021)
7. W. Jin, Z. Zhao, Y. Li, J. Li, J. Xiao, Y. Zhuang, Video question answering via knowledge-based progressive spatial-temporal attention network. ACM Trans. Multimedia Comput. Commun. Appl. (TOMM) **15**(2s), 1–22 (2019)
8. J. Kim, M. Ma, K. Kim, S. Kim, C.D. Yoo, Progressive attention memory network for movie story question answering, in *Proceedings of the IEEE/CVF Conference on Computer Vision and Pattern Recognition* (2019), pp. 8337–8346
9. K.-M. Kim, M.-O. Heo, S.-H. Choi, B.-T. Zhang, Deepstory: video story QA by deep embedded memory networks. arXiv preprint arXiv:1707.00836 (2017)
10. T.M. Le, V. Le, S. Venkatesh, T. Tran, Learning to reason with relational video representation for question answering. arXiv preprint arXiv:1907.04553, 2 (2019)
11. T.M. Le, V. Le, S. Venkatesh, T. Tran, Hierarchical conditional relation networks for video question answering, in *Proceedings of the IEEE/CVF Conference on Computer Vision and Pattern Recognition* (2020), pp. 9972–9981
12. X. Li, J. Song, L. Gao, X. Liu, W. Huang, X. He, C. Gan, Beyond rnns: positional self-attention with co-attention for video question answering, in *Proceedings of the AAAI Conference on Artificial Intelligence*, vol. 33 (2019), pp. 8658–8665
13. J. Mun, P. Hongsuck Seo, I. Jung, B. Han, Marioqa: answering questions by watching gameplay videos, in *Proceedings of the IEEE International Conference on Computer Vision* (2017), pp. 2867–2875

14. S. Na, S. Lee, J. Kim, G. Kim, A read-write memory network for movie story understanding, in *Proceedings of the IEEE International Conference on Computer Vision* (2017), pp. 677–685
15. X. Song, Y. Shi, X. Chen, Y. Han, Explore multi-step reasoning in video question answering, in *Proceedings of the 26th ACM International Conference on Multimedia* (2018), pp. 239–247
16. A. Vaswani, N. Shazeer, N. Parmar, J. Uszkoreit, L. Jones, A.N. Gomez, Ł. Kaiser, I. Polosukhin, Attention is all you need, in *Advances in Neural Information Processing Systems* (2017), pp. 5998–6008
17. D. Xu, Z. Zhao, J. Xiao, F. Wu, H. Zhang, X. He, Y. Zhuang, Video question answering via gradually refined attention over appearance and motion, in *Proceedings of the 25th ACM International Conference on Multimedia* (2017), pp. 1645–1653
18. H. Xue, W. Chu, Z. Zhao, D. Cai, A better way to attend: attention with trees for video question answering. IEEE Trans. Image Process. **27**(11), 5563–5574 (2018)
19. J. Zhang, Y. Peng, Object-aware aggregation with bidirectional temporal graph for video captioning, in *Proceedings of the IEEE/CVF Conference on Computer Vision and Pattern Recognition* (2019), pp. 8327–8336

Part IV
Advanced Topics in VQA

In addition to traditional, classic VQA problems, which are aimed at answering a natural language question regarding an image or video, many advanced tasks can be derived from VQA. In this part, we introduce advanced topics related to VQA. Several of these aspects are based on different input domains, such as embodied VQA, medical VQA and text-based VQA. Other topics pertain to different tasks, such as visual question generation, visual dialog and referring expression.

Chapter 10
Embodied VQA

Abstract It is a long-standing goal for scientists to develop robots that can perceive, communicate with humans in natural language and complete commands as requested. Several sub-tasks are proposed to achieve this goal in sequential manner, e.g. Vision-and-Language Navigation requires the intelligent agent to follow detailed instructions with visual perception, Remote object localization gives the agent shorter and more abstract instructions, Embodied QA expects the agent to actively explore the environment and respond to inquiries, Interactive QA hopes the agent actively interact with a virtual environment to get responses of inquiries. In this chapter, we first briefly introduce some mainstream simulators, datasets and evaluation criteria that benchmark applied in this field, such as MatterPort3D, iGibison and Habitat et al. Subsequently, we describe the motivation, methodology and key performance of several methods corresponding to each sub-tasks.

10.1 Introduction

Scientists have made unremitting attempts to build an intelligent agent that can actively perceive the environment through vision, audition and others sensors, communicate with users via natural language, and act in virtual or even real scenarios. With the considerable research on classical visual question answering (VQA), both the research communities of computer vision and natural language processing have paid more attention to embodied visual question answering (Embodied VQA).

Visual-and-language tasks for embodied robotics and VQA involve certain similarities. For both tasks, the core scientific problem is multimodal information alignment. Notably, the VQA task combines two modalities, i.e. vision and natural language, whereas Embodied VQA task combines three modalities, i.e. vision, natural language and actions. In Embodied VQA tasks, intelligent agents are presented with virtual or real environments and natural language instructions, then the agents must respond to the instruction by actively exploring in the environment, thereby obtaining much more visual information comparing with traditional VQA tasks. According to the different levels of the tasks, the natural language instruction may be specific and detailed [1] or abstract and concise [16]. To be more specific, the instruction can be

© The Author(s), under exclusive license to Springer Nature Singapore Pte Ltd. 2022
Q. Wu et al., *Visual Question Answering*, Advances in Computer Vision and Pattern Recognition, https://doi.org/10.1007/978-981-19-0964-1_10

as detailed as *"Go straight along the corridor, and turn left before the white table"* or simply like *"how many white chairs are there in this house?"*. Intelligent agents follow the abovementioned instruction and reach the expected destination or explore the whole environment to determine what the answer is.

In this chapter, we present a comprehensive review of Embodied VQA considering three categories based on the nature of skills required. According to the degree of difficulty, it can be divided into the following three progressive tasks: language-guided visual navigation, embodied question answering and interactive question answering.

- **Language-guided Visual Navigation**. The language-guided visual navigation task aims to enable intelligent agents to follow natural language instruction, combine visual input from the environment and move to the expected spot. Furthermore, this task can be divided into two sub-tasks: vision-and-language navigation (VLN) and remote object localization (ROL).

- **Embodied Question Answering (EQA)**. Based on language-guided visual navigation, the embodied question answering task requires intelligent agents to actively explore an unknown environment, navigate autonomously and give a response to the question asked.

- **Interactive Question Answering (IQA)**. Interactive question answering is similar to an advanced version of embodied question answering, but requires agents to interact with an unknown environment.

Furthermore, in this chapter, we describe the datasets available for each task, corresponding evaluation parameters and commonly adopted simulators/platforms. The datasets vary widely along two dimensions: (i) their size, i.e. the number of paths and natural language instructions, and (ii) the environment, i.e. virtual or photo-realistic and indoor or outdoor environment.

10.2 Simulators, Datasets and Evaluation Criteria

A number of simulators, datasets and evaluation criteria have been proposed for each task. Simulators provide a virtual environment for intelligent agents, and thus similar simulators are usually used. However, the datasets and evaluation criteria vary significantly due to the uniqueness of the associated task requirements.

10.2.1 Simulators

Simulators (platforms) are employed to ensure that intelligent agents can cruise, implement actions and obtain feedback information via certain APIs. Commonly used simulators include MatterPort3D [1], House3D [20], Habitat [17], AI2-THOR [10], CHALET [21] and iGibson [14, 18] et al. The main differences include the vision style, interactivity and continuity (whether the intelligent agent can move

Table 10.1 Major simulators for Embodied VQA and their main characteristics

Environment	Navigable	3D Scene Scans	3D Asset Library	Physics-Based Interaction	Object States	Object Specific Reactions	Dynamic Lighting	Multiple Agents	Real Counterpart
AI2-THOR	✓		✓	✓	✓	✓	✓	✓	✓
iGibson	✓	✓		✓				✓	
Habitat	✓	✓		Collisions					
Matterport3D	✓	✓							
Minos	✓	✓							

to any reachable/navigable points). Details of these simulators are presented in the following text, and the key characteristics are summarized in Table 10.1.

MatterPort3D

The MatterPort3D simulator [1] is based on the MatterPort3D dataset [2], which is a large-scale RGB-D dataset containing 10,800 panoramic views from 194,400 RGB-D images of 90 building-scale scenes. The MatterPort3D dataset also includes depth, camera pose and 2D and 3D semantic segmentations. The MatterPort3D simulator uses the MatterPort3D dataset as the source of photo-realistic visual data and enables intelligent agents to observe horizontal 360 degrees and pitch $[0\text{-}2\pi)$ RGB images at certain points. Intelligent agents travel by selecting a new viewpoint, nominating camera heading and adjusting elevation. Notably, the MatterPort3D simulator prepares a navigation graph to illustrate the connectivity between each viewpoint in advance.

House3D

House3D [20] is a virtual 3D environment that consists of over 45k indoor scenes equipped with a diverse set of scene types, layouts and objects sourced from the SUNCG dataset. All 3D objects are fully annotated with category labels. Agents in the environment have access to observations of multiple modalities, including RGB images, depth, segmentation masks and top-down 2D map views.

Habitat

Habitat [17] is a photo-realistic simulation environment based on the MatterPort3D, Replica and 2D-3D-S datasets, providing real-time rendered RGB, RGB-D and depth data. Remarkably, Habitat offers a continuous simulation environment and fast rendering performance (over 10,000 fps multi-process on a single GPU). Thus, this environment is commonly employed in continuous environment language-guided visual navigation tasks.

AI2-THOR

AI2-THOR [10] is an interactive 3D environment that consists of internet vision-style 3D indoor scenes and interactable objects. The scenes are manually reconstructed

Fig. 10.1 Types of questions included in the EQA dataset

	location:	*'What room is the <OBJ> located in?'*
	color:	*'What color is the <OBJ>?'*
EQA v1	color_room:	*'What color is the <OBJ> in the <ROOM>?'*
	preposition:	*'What is <on/above/below/next-to> the <OBJ> in the <ROOM>?'*

by artists from the reference photo, and thus AI2-THOR does not have a bias that usually exists in scenes generated automatically.

CHALET

CHALET [21] is an internet vision-style 3D indoor scenario simulator with 58 rooms and 10 houses. As an interactable environment, CHALET provides a range of common household activities, like moving objects, toggling appliances and placing objects inside closable containers.

iGibson

iGibson [14, 18] is an internet vision-style interactive 3D indoor simulation environment containing 15 home-sized scenes with 108 rooms. As a replica of real-world homes, the environment does not involve the bias that exists in automatically generated environments. In addition to RGB-D images, the environment provides depth, segmentation, LiDAR and flow data.

In a nutshell, specific simulators are suitable for certain tasks. MatterPort3D and Habitat are commonly used in VLN tasks since both frameworks provide a photorealistic simulation environment. House3D is employed in the Embodied QA task, while iGibson and AI2-THOR are usually used in the interactive QA task due to their interactivity capability.

10.2.2   Datasets

The datasets for different tasks vary significantly. In this section, we are going to introduce several datasets.

Commonly used datasets for vision-and-language navigation task include R2R, RxR, Habitat and REVERE et al.

The R2R dataset [1] is based on the MatterPort3D dataset. The dataset contains 21,567 open-vocabulary crowd-sourced navigation instructions with an average length of 29 words. Each instruction describes the method to navigate from the starting point to the corresponding destination. The whole dataset is divided into training, validation and testing parts. Derivatively, a fine-grained R2R dataset [7] is proposed, augmenting the original R2R dataset by adding sub-instructions. However, despite the significant meaning of R2R dataset, which means that it is the first dataset for VLN task, it works on a discrete environment with only indoor scenarios. The presence of navigation graphs is not suitable for real application settings.

Environments	Unique Questions	Total Questions
train 643	147	4246
val 67	104	506
test 57	105	529

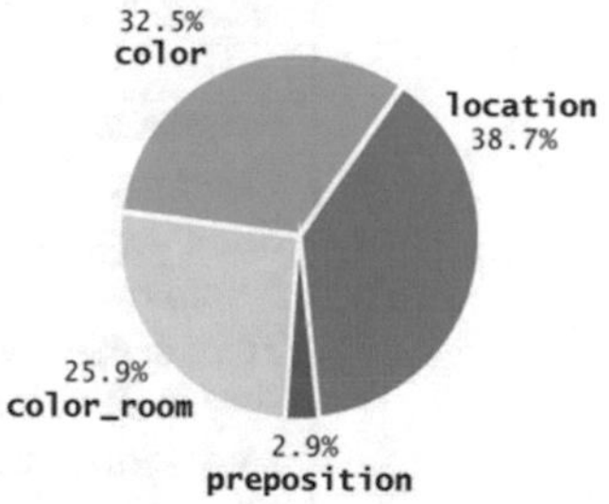

Fig. 10.2 Overview of the EQA v1 dataset including dataset split statistics (left) and question type breakdown (right)

The RxR dataset [12] is proposed for the VLN task. Compared with the first dataset proposed for the VLN task, the RxR dataset has a (i) significantly larger scale: the dataset contains more than 126K paths and corresponding instructions; (ii) finer grain: during the annotation process, the annotator must move and provide instructions by talking simultaneously, therefore, alignment in time and space can be attained between the instructions, visual perception and actions.

The Habitat dataset [17] is proposed for the VLN task in a continuous environment. Based on the original R2R dataset, Habitat rearranged the action space and rebuilt the form of paths of R2R.

The EQA dataset [3] is proposed for Embodied QA tasks. The dataset is based on the House3D simulator [20] and CLEVR dataset [8] to build a virtual environment and generate grounded questions and answers. The dataset classifies questions into certain types, as shown in Fig. 10.1. The EQA dataset contains over 5,000 questions across more than 750 environments, referring to 45 unique objects in 7 unique room types. The split statistics and question type breakdown of this dataset are shown in Fig. 10.2.

The interactive question/answer dataset (IQUAD) V1 dataset [5] is proposed for the interactive QA task. The AI2-THOR simulator has over 75,000 multiple choice questions and corresponding answers. Similar to the EQA dataset, the IQUAD dataset generates questions in a variety of types, and the overview is shown in Fig. 10.3.

10.2.3   Evaluations

The evaluation metrics for specific tasks are also significantly different.

For the language-guided visual navigation task, the following aspects must be considered: (i) measurement from the stopping point to the expected destination, (ii) measurement of the similarity of paths and (iii) reasonable penalty for longer paths even if the destination is reached. Thus, the following evaluation metrics are usually used: path length (PL), navigation error (NE), success rate (SR), success

Interactive Question Answering Dataset Statistics		
	Train	Test
Existence	25,600	640
Counting	25,600	640
Spatial Relationships	25,600	640
Rooms	25	5
Total scene configurations (s.c.)	76,800	1,920
Avg # objects per (s.c.)	46	41
Avg # interactable objects (s.c.)	21	16
Vocabulary Size	70	70

Fig. 10.3 Statistics of the IQUAD dataset: it has a variety of question types, objects and scene configurations

weighted by path length (SPL), coverage weighted by length score (CLS) normalized dynamic time warping (NDTW) and success weighted by normalized dynamic time warping (SDTW) et al.

Similar to language-guided visual navigation, Embodied QA has the following evaluation metrics: (i) distance to the target object at navigation termination, (ii) change in distance to target from initial to final position, (iii) shortest distance to the target at any point in the episode, (iv) percentage of questions for which an agent either terminates in or enters the room containing the target object(s) and (v) percentage of episodes in which agents choose to terminate navigation and answer before reaching the maximum episode length.

Similar to the Embodied QA task, the interactive question answering task has the following evaluation metrics: (i) answer accuracy, (ii) path length and (iii) percentage of invalid action et al.

10.3 Language-Guided Visual Navigation

It is a long-standing goal for AI researchers to enable intelligent agents to navigate to the expected destination point with human natural language instructions and environmental image/video stream input. Language-guided visual navigation tasks can be divided into two tasks according to the specificity of the natural language instructions given, i.e. vision-and-language navigation (VLN) and remote object localization (ROL), which are introduced in the following sections.

10.3.1 Vision-and-Language Navigation

The vision-and-language navigation (VLN) task requires intelligent agents to listen to general, verbal instructions and navigate in the virtual environment according to the provided instructions. During the navigation process, intelligent agents must combine the information from both vision and natural language for analysis, and then take actions to move around the environment and acquire new information, repeat such process until they arrive at the predetermined destinations. Thus, the core scientific problem of vision-and-language navigation is cross-modal information alignment. In contrast to the usual VQA task, which usually involves two modalities (vision and language), the VLN task requires information alignment between three modalities: vision, natural language and action.

Generally, the methodologies for vision-and-language navigation can be classified into three paradigms associated with imitation learning, reinforcement learning and self-supervision. Furthermore, with the development of the VLN task, recent researches on the VLN task have attempted to combine these paradigms.

10.3.1.1 Imitation Learning Methodology

Motivation

The imitation learning paradigm is a key methodology to solve the VLN task in the early stage. The basic logic is that intelligent agents learn the strategy of making decisions according to the existing decision and behavior data provided by human experts. This process is also known as behavior cloning. In the VLN task, the intelligent agents extract features from the decision data (states and action sequences) provided by the human experts and evolve the optimal strategy model.

Method

Anderson et al. [1] proposed a recurrent neural network policy using an LSTM-based sequence-to-sequence architecture with an attention mechanism for intelligent agents. The intelligent agent considers the current image and the previous action as the encoder input of the model, applies attention to the hidden state of language encounters and predicts the distribution over the next action. Fried et al. [4] employed a speaker-follower model to increase the navigation success rate and noted that when implementing follower data argumentation, the performance will be improved by more than two times that for the previous benchmark. The authors employed ground-truth routes and annotation descriptions to train the speaker and used followers to synthesize instructions to add into the original dataset for argument, thereby accelerating the training process. The model framework is shown in Fig. 10.4. Recently, Hong et al. [7] employed sub-instruction attention and shifting modules to increase the navigation success rate.

Performance and Limitation

The imitation learning methodology provides the most direct and simple solution for the VLN task. This method represents the first benchmark for future academic investigation. However, the framework involves a notable limitation. The experts could and only could sample limited pairs of observations and instructions. If the

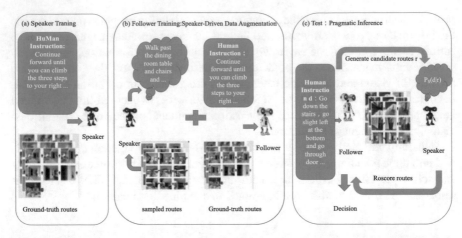

Fig. 10.4 The speaker-follower model combines an instruction follower model and a speaker model. **a** The speaker model is trained on ground-truth routes with human-generated descriptions; **b** the model provides the follower with additional synthetic instruction data to realize bootstrap training; **c** the model helps the follower interpret ambiguous instructions and choose the best route during inference

agent encounters some cases that don't appear in the dataset, it may be lost and at a loss. In other words, the agent will just simply copy every experts' behavior, even irrelevant actions through imitation learning because this method takes all errors equally. Such agents are not intelligent enough in practical use.

10.3.1.2   Reinforcement Learning Paradigm

Motivation

The reinforcement learning (RL) methodology is another key methodology for solving vision-and-language navigation tasks. The main logic is that by interacting with the virtual environment and setting a proper reward system, intelligent agents autonomously explore the environment and learn navigation strategies. Most of the imitation learning models for VLN tasks fail to address generalization problems due to the considerable gap between behavior cloning and real-world practices. The introduction of reinforcement learning is vital for further improvement of VLN models.

Method

Wang et al. [19] proposed a planned-ahead hybrid reinforcement learning model to solve the generalization problem. In this framework, considering the sequential decision-making nature of the VLN task, a reinforced planning head (RPA) is employed, as shown in Fig. 10.5. The RPA architecture consists of model-free and model-based paths. The model-based path consists of multiple look-ahead modules and one aggregation module. At each step, the recurrent policy model considers the

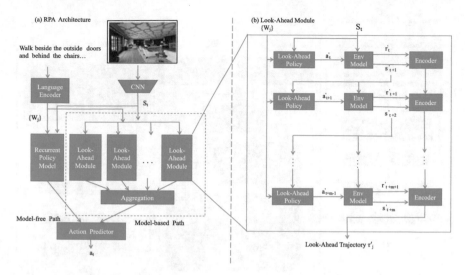

Fig. 10.5 Overview of the PRA architecture

word features and states as input and yields information regarding the next procedure (predicted action). Notably, the model-based path framework only predicts the
potential action, and subsequently, the action predictor chooses the final action based
on the information from the model-free and model based paths.

Performance and Limitation

The proposed model outperforms the existing benchmark on validation of unseen
datasets.

Lansing et al. [13] established a visual language navigation model based on the
RL framework for dialog instructions. This model was successfully lightweight and
applied to indoor navigation scenarios.

10.3.1.3 Self-Supervised Learning Paradigm

Motivation

Self-supervised methodology is the third key methodology to solve the vision-and-
language navigation task. The main logic is that intelligent agents learn from semi-
human expert behavior according to certain algorithms and evolve to the optimal strategy policy. Unlike the paradigm of imitation learning, self-supervised methodology
requires the algorithm to produce certain labels, in this case, instructions or routers.
In this domain, researchers have started to combine the self-supervised methodology with imitation learning and reinforcement learning methodology as a practical
methodology to enhance the navigation ability of intelligent agents in validation
unseen environments. Notably, the previous models fail to follow instructions prop-

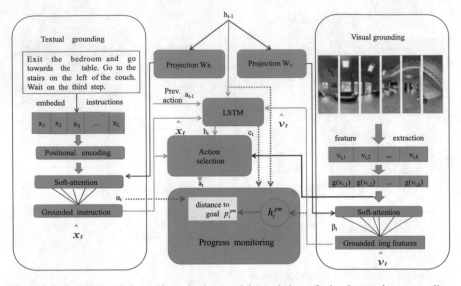

Fig. 10.6 Overview of the self-monitoring model consisting of visual-textual co-grounding, progress monitoring and action selection modules. Textual grounding: identify the part of the instruction that has been completed or is in progress and the part that is potentially required for subsequent action. Visual grounding: summarize the observed surrounding images. Progress monitor: regularize and ensure that the grounded instruction reflects progress toward the goal. Action selection: identify the direction in which to proceed

erly; thus, although the agents may reach the endpoint, it is necessary to monitor the navigation progress.

Method

Ma et al. [15] introduced a self-monitoring agent with two complementary components: a visual-textual co-grounding module and a progress monitor. A progress monitor serves as a regularizer and estimates the navigation process by conditioning on three inputs: history of grounded images and instructions, current observation of the surrounding images and positions of grounded instructions (Fig. 10.6).

Performance and Limitation

The self-monitoring method exhibits a significantly high success rate (more than 8% on an unseen test set). However, this framework is based on a discrete environment and limited to indoor scenes and thus cannot be applied in real-world application settings.

10.3.1.4   New Era: VLN in Continuous Environment

Motivation

The existing approaches to vision-and-language navigation rely on the R2R dataset proposed by Anderson et al. [1]. However, in real-world application scenarios, a navigation graph may not be available in advance, and intelligent agents must explore any point reachable in an unknown environment that they are exposed to. In addition, a panoramic picture is not provided in a real-world scenario, and the intelligent agent is provided with only the first-person view (FPV). Thus, it is necessary to rebuild the action space, which can be used in continuous environment navigation with the provided FPV.

Method

Krantz et al. [11] eliminated the existing navigation graph (topology) used in the R2R dataset and proposed a novel action space. This framework regulates certain actions in advance, including "turn left", "turn right" and "move forward". The action of turning left/right actually means turning left by 15 degrees and that of moving forward means advancing 0.25 m. For example, if the intelligent agent predicts its next action as turning left 45 degrees, the simulator converts this predicted action into three instances of the "turn left" action. In this manner, the new vision-and-language task, vision-and-language navigation in a continuous environment (VLN-CE), has been defined.

Models for the VLN-CE task include sequence-to-sequence and cross-modal attention, as shown in Fig. 10.7. The basic architecture of the sequence-to-sequence model is similar to that of the imitation learning methodology and is thus not repeated. The proposed cross-model attention model consists of two recurrent networks: one network tracks visual observations and the other network makes decisions based on attended instructions and visual features.

Performance and Limitation

Nearly all evaluation metrics of VLN-CE are considerably inferior to those of the VLN model for discrete environments (R2R dataset) since the task setting leads to additional action steps being required to accomplish the same action. Although research on high-level instructions still remains limited, Krantz et al. contributed to the development of VLN tasks toward real-world applications.

10.3.2   Remote Object Localization

Motivation

A 10-year-old child can easily follow the instruction "fetch me a pillow", even in a completely unknown environment. However, it is fairly difficult for robots to accomplish such a task because robots have difficulty in learning knowledge from an environment that they have explored and transferring it suitably when encountering a

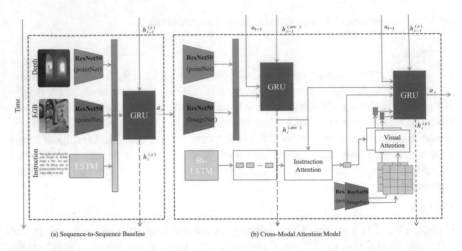

Fig. 10.7   Models for the VLN-CE task

new environment. For example, a pillow usually appears on a sofa, which usually appears in a living room, and the living room is usually connected with another room via a corridor. In addition, humans can understand high-level natural language instructions and connect them with visual perception. To enable robots to exhibit this ability and interact with human beings in a more flexible and accurate manner, the remote embedded visual preferring expression task has been established.

In this task, the intelligent agent is placed in a random position, provided an instruction relevant to a remote object, for example, "bring me the bottom picture that is next to the top of stairs on level one", and the robot has to explore and find the target object according to the instruction and visual images perceived. However, the referred target remote objects may not always be directly visible. In this scenario, intelligent agents must have the common-sense reasoning ability to reach the right position in which the remote object might be found.

Method

Qi et al. [16] introduced an interactive navigator-pointer model, as shown in Fig. 10.8. The model consists of pointer instruction and navigation sub-modules. The pointer (object localization) module includes a local visual perception image and natural language instruction as the input and returns three objects that best fit the instructions. The visual features and labels of the three obtained objects are the input of the navigator module. Furthermore, the navigator module employs natural language instruction and a perceived image of the current position as the input, and specifies a stop signal or direction of the next step. If the navigator stops outputting, the suitable object that the current pointer returns is considered the final result. This framework adopts the FAST algorithm [9] as the navigator module and MAttNet [23] as the point module.

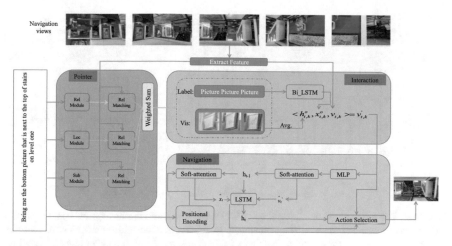

Fig. 10.8 Overview of interactive navigator-pointer model

Performance and Limitation

The success rate of the random algorithm is less than 1% and R2R-TF and R2R-SF exhibit a success rate of 2% in unknown environments. The proposed algorithm has a success rate of more than 11% for validation in this environment. However, there remains a significant gap with the success rate of humans, 77.84%, which highlights the scope of further work in this domain.

10.4 Embodied QA

Motivation

To build an intelligent agent that can perceive its environment, researchers in both robotics and computer science domains have attempted to realize communication with humans via natural language and implement actions in a real-world environment. Das et al. [3] presented a new task named embodied question answering, in which an agent is spawned at a random location in a 3D environment and is asked a question ('What color is the car?'). To answer this query, the agent must first intelligently navigate to explore the environment, gather necessary visual information through first-person (egocentric) vision, and answer the question ('orange').

Compared to traditional language-guided visual navigation task, Embodied QA task requires more of an intelligent agent model. A typical instance is an active perception: the VLN task requires the intelligent agent to follow the instruction given by humans, while in the Embodied QA task, the intelligent agent is required to actively explore the virtual environment to determine the answer to the question. Second, the question presented to the intelligent agent is a high-level natural language instruc-

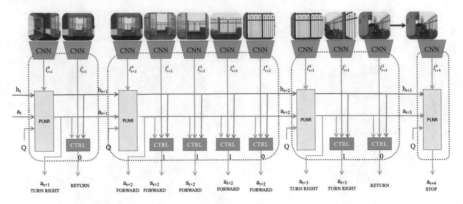

Fig. 10.9 The adaptive computation time (ACT) navigator splits the navigation task between a planner and a controller module. The planner selects actions, and the controller decides to continue performing that action for a variable number of time steps, resulting in a decoupling of direction ('turn left') and velocity ('5 times') and strengthening of the long-term gradient flows of the planner module

tion. In this scenario, the intelligent agent must realize common-sense processing and reasoning during the execution of task.

Method

Das et al. [3] introduced a hierarchical model for Embodied QA. Because the model for Embodied QA involves four forms of information: vision, language, navigation and answer, four separate nature modules are employed. In particular, different forms of information demand corresponding neural network architectures, and thus, in this framework, convolutional neural networks (CNNs) and recurrent neural networks (RNNs) are employed. Because of the significant variance between different questions, the navigation steps for solving the questions are dramatically different. Adaptive computation time (ACT) RNNs proposed by Graves [6] are used as the basic framework of the planner and controller, as shown in Fig. 10.9.

A pretrained CNN network and two layers of LSTMs with 128-dimensional hidden states are used for vision-and-language encoding, respectively.

Furthermore, to increase the efficiency of intelligent agents in finding the answer to questions, Yu et al. [22] proposed a generalized task of Embodied QA, named multi-target embodied question answer. This task is aimed at examining complicated questions that can be decomposed into two or more meta-questions, such as "Is the dresser in the bedroom larger than the oven in the kitchen?" Multi-target Embodied QA has an architecture similar to that of the Embodied QA (Fig. 10.10). RNNs are used as navigation and controller modules, and convolutional neural networks are used as future extractors for visual modules. Unlike traditional Embodied QA, the questions in which are relatively simple, multi-target Embodied QA involves more complicated questions. Thus, a program generator has been developed to decode the structural information from complicated questions.

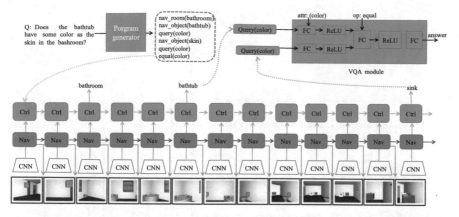

Fig. 10.10 Model architecture: The proposed model is composed of a program generator, navigator, controller and VQA module

Performance and Limitation

Both approaches are of significance to the vision-and-language community, although the success rate and accuracy of navigation are not satisfactory. In addition, the questions asked are relatively simple and structural, and in real-world application scenarios, the variability of the question form cannot be neglected. Thus, additional research must be performed on embodied questions and answers.

10.5 Interactive QA

Motivation

AI researchers aim to create intelligent agents that can perform manual tasks in real-world scenarios and communicate with human beings via natural language. In contrast to the Embodied QA task, interactive question answering requires intelligent agents to interact with the virtual environment and manipulate objects in the environment, for example, "open the microwave oven", "open the refrigerator door" or move an object at a location to answer the questions.

Method

Gordon et al. [5] introduced a hierarchical interactive memory network (HIMN), as shown in Fig. 10.11. This model decomposes a complicated task into multiple sub-tasks to reduce the complexity of each task, allowing the system to operate, learn and reason across multiple time scales. Specifically, the planner is designed to choose the task to be performed, such as navigation, manipulation, answering and presenting the specific command. Subsequently, the task is conducted using a set of low-level controllers, including a navigator, manipulator, detector, scanner and

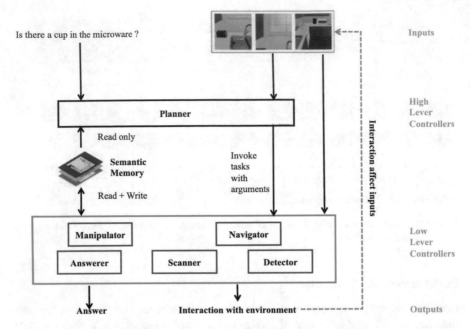

Fig. 10.11 Overview of the hierarchical interactive memory network (HIMN)

answerer. When the task or the sub-task is terminated, the controller returns control to the planner. Therefore, when the model faces certain independent tasks, the tasks can be conducted independently.

Performance and Limitation

This interactive question and answer task promoted the application of vision-and-language cross-modal models in the robotics domain. However, this approach involves the following limitations: first, due to the 2D nature of the segmentation map, the proposed model can not differentiate whether an object is inside a container or on top of it. Second, the proposed model is fairly ineffective at exploring the environment. Third, the virtual environment is not photo-realistic, and there exists a nonneglectable domain gap when the internet vision-style vision input is transferred to a photo-realistic input.

References

1. P. Anderson, Q. Wu, D. Teney, J. Bruce, M. Johnson, N. Sünderhauf, I. Reid, S. Gould, A. Van Den Hengel, Vision-and-language navigation: interpreting visually-grounded navigation instructions in real environments, in *Proceedings of the IEEE Conference on Computer Vision and Pattern Recognition* (2018), pp. 3674–3683

2. A. Chang, A. Dai, T. Funkhouser, M. Halber, M. Niessner, M. Savva, S. Song, A. Zeng, Y. Zhang, Matterport3D: learning from RGB-D data in indoor environments, in *International Conference on 3D Vision (3DV)* (2017)
3. A. Das, S. Datta, G. Gkioxari, S. Lee, D. Parikh, D. Batra, Embodied question answering, in *Proceedings of the IEEE Conference on Computer Vision and Pattern Recognition* (2018), pp. 1–10
4. D. Fried, R. Hu, V. Cirik, A. Rohrbach, J. Andreas, L.-P. Morency, T. Berg-Kirkpatrick, K. Saenko, D. Klein, T. Darrell, Speaker-follower models for vision-and-language navigation. *arXiv preprint* arXiv:1806.02724 (2018)
5. D. Gordon, A. Kembhavi, M. Rastegari, J. Redmon, D. Fox, A. Farhadi, Iqa: visual question answering in interactive environments, in *Proceedings of the IEEE Conference on Computer Vision and Pattern Recognition* (2018), pp. 4089–4098
6. A. Graves, Adaptive computation time for recurrent neural networks. *arXiv preprint* arXiv:1603.08983 (2016)
7. Y. Hong, C. Rodriguez-Opazo, Q. Wu, S. Gould, Sub-instruction aware vision-and-language navigation. *arXiv preprint* arXiv:2004.02707 (2020)
8. J. Johnson, B. Hariharan, L. Van Der Maaten, L. Fei-Fei, C. Lawrence Zitnick, R. Girshick, Clevr: a diagnostic dataset for compositional language and elementary visual reasoning, in *Proceedings of the IEEE Conference on Computer Vision and Pattern Recognition* (2017), pp. 2901–2910
9. L. Ke, X. Li, Y. Bisk, A. Holtzman, Z. Gan, J. Liu, J. Gao, Y. Choi, S. Srinivasa, Tactical rewind: self-correction via backtracking in vision-and-language navigation, in *Proceedings of the IEEE/CVF Conference on Computer Vision and Pattern Recognition* (2019), pp. 6741–6749
10. E. Kolve, R. Mottaghi, W. Han, E. VanderBilt, L. Weihs, A. Herrasti, D. Gordon, Y. Zhu, A. Gupta, A. Farhadi, Ai2-thor: an interactive 3d environment for visual AI. *arXiv preprint* arXiv:1712.05474 (2017)
11. J. Krantz, E. Wijmans, A. Majumdar, D. Batra, S. Lee, Beyond the nav-graph: vision-and-language navigation in continuous environments, in *European Conference on Computer Vision* (Springer, 2020), pp. 104–120
12. A. Ku, P. Anderson, R. Patel, E. Ie, J. Baldridge, Room-Across-Room: multilingual vision-and-language navigation with dense spatiotemporal grounding, in *Conference on Empirical Methods for Natural Language Processing (EMNLP)* (2020)
13. L. Lansing, V. Jain, H. Mehta, H. Huang, E. Ie, Valan: vision and language agent navigation. *arXiv preprint* arXiv:1912.03241 (2019)
14. C. Li, F. Xia, R. Martín-Martín, M. Lingelbach, S. Srivastava, B. Shen, K. Vainio, C. Gokmen, G. Dharan, T. Jain, A. Kurenkov, K. Liu, H. Gweon, J. Wu, L. Fei-Fei, S. Savarese, igibson 2.0: object-centric simulation for robot learning of everyday household tasks (2021)
15. C.-Y. Ma, J. Lu, Z. Wu, G. AlRegib, Z. Kira, R. Socher, C. Xiong, Self-monitoring navigation agent via auxiliary progress estimation. *arXiv preprint* arXiv:1901.03035 (2019)
16. Y. Qi, Q. Wu, P. Anderson, X. Wang, W. Y. Wang, C. Shen, A. v. d. Hengel, Reverie: Remote embodied visual referring expression in real indoor environments, in *Proceedings of the IEEE/CVF Conference on Computer Vision and Pattern Recognition* (2020), pp. 9982–9991
17. M. Savva, A. Kadian, O. Maksymets, Y. Zhao, E. Wijmans, B. Jain, J. Straub, J. Liu, V. Koltun, J. Malik, D. Parikh, D. Batra, Habitat: a platform for embodied AI research, in *Proceedings of the IEEE/CVF International Conference on Computer Vision (ICCV)* (2019)
18. B. Shen, F. Xia, C. Li, R. Martín-Martín, L. Fan, G. Wang, C. Pérez-D'Arpino, S. Buch, S. Srivastava, L.P. Tchapmi, M.E. Tchapmi, K. Vainio, J. Wong, L. Fei-Fei, S. Savarese, igibson 1.0: a simulation environment for interactive tasks in large realistic scenes (2021)
19. X. Wang, W. Xiong, H. Wang, W.Y. Wang, Look before you leap: bridging model-free and model-based reinforcement learning for planned-ahead vision-and-language navigation, in *Proceedings of the European Conference on Computer Vision (ECCV)* (2018), pp. 37–53
20. Y. Wu, Y. Wu, G. Gkioxari, Y. Tian, Building generalizable agents with a realistic and rich 3d environment. *arXiv preprint* arXiv:1801.02209 (2018)

21. C. Yan, D. Misra, A. Bennnett, A. Walsman, Y. Bisk, Y. Artzi, Chalet: Cornell house agent learning environment. *arXiv preprint* arXiv:1801.07357 (2018)
22. L. Yu, X. Chen, G. Gkioxari, M. Bansal, T.L. Berg, D. Batra, Multi-target embodied question answering, in *Proceedings of the IEEE/CVF Conference on Computer Vision and Pattern Recognition* (2019), pp. 6309–6318
23. L. Yu, Z. Lin, X. Shen, J. Yang, X. Lu, M. Bansal, T.L. Berg, Mattnet: modular attention network for referring expression comprehension, in *Proceedings of the IEEE Conference on Computer Vision and Pattern Recognition* (2018), pp. 1307–1315

Chapter 11
Medical VQA

Abstract Inspired by the rise of VQA research in general domain, the task of Medical VQA has received great attention from computer vision, natural language processing and biomedical research communities in recent years. Given a medical image and clinically related question about the visual elements in the medical image, a Medical VQA system is required to deeply comprehend both the medical image and the asked question to predict the correct answer. In this chapter, we first introduce mainstream datasets used for Medical VQA tasks, such as VQA-RAD, VQA-Med, PathVQA and SLAKE datasets. Then, we elaborate the prevalent methods for Medical VQA tasks in detail. These methods can be classified into three categories based on their main characteristics: classical VQA methods, meta-learning methods and BERT-based methods for Medical VQA.

11.1 Introduction

Medical images play a vital role in clinical diagnosis and treatment, but the need for diagnosis and the reporting of image-based examinations considerably exceeds the current medical capabilities of physicians. Recently, a number of computer-assisted medical diagnosis technologies have been proposed to help relieve the pressure of healthcare systems. In resource-limited conditions, the medical VQA task can provide a "second opinion" to radiologists regarding their analysis of the image, and the responses can be used by patients to obtain basic information regarding the medical image, without consulting their doctor. As a domain-specific branch of general VQA task, the Medical VQA task is performed by inputting medical images accompanied by clinical-related questions, with the system being expected to correctly answer these clinical questions in natural language according to the visual clues in the medical images.

In this chapter, we first review six mainstream datasets specifically proposed for Medical VQA tasks, namely, VQA-Med-2018 [8], VQA-Med-2019 [3], VQA-Med-2020 [1], VQA-RAD [13], PathVQA [9] and SLAKE [14]. We compare the similarity and difference between these datasets and give detailed descriptions of each dataset.

Then, we comprehensively present the review of Medical VQA methods that are classified into three categories according to their main characteristics and contributions as classical VQA methods, meta-learning methods and vision-language pre-training methods for Medical VQA.

First, *classical VQA methods* (Sect. 11.3) for Medical VQA are motivated by the classical methods proposed for general VQA tasks. These methods usually utilize convolution networks (CNNs) such as VGGNet and ResNet networks to learn the embedding of medical images, recurrent networks (RNNs) such as LSTM, Bi-LSTM networks to learn the embedding of clinical questions and classical feature fusion strategies such as joint embedding and attention mechanisms to learn the fused multi-modal features, a multi-layer classifier or sequence-to-sequence encoder-decoder to predict the answer as classification task or generation task.

Second, *meta-learning methods* (Sect. 11.4) for Medical VQA utilize meta-learning to overcome the severe labeled data limitation in Medical VQA. Rather than using CNN networks pretrained on ImageNet to learn visual features from a limited number of medical images, these methods train meta-models directly on medical images, of which the weights can be adapted to Medical VQA tasks more easily than weights of CNNs pretrained on ImageNet.

Third, *BERT-based methods* (Sect. 11.5) for Medical VQA are inspired by the successful application of BERT and vision-language pre-training in general domain. Most BERT-based methods for Medical VQA simply utilize BERT as language encoder to learn textual features from clinical questions and share similar fusion architecture with classical methods. Other work may use Transformer to interact between the two modalities. And more recently, some work pre-train BERT-like models on data of medical image-text pairs and finetune these models on several Medical VQA datasets.

11.2   Datasets

Compared with the classical VQA task that focuses on general domain, Medical VQA is specifically designed for answering clinical-related questions based on the visual elements in the given medical images. In order to achieve this goal, specialized VQA datasets that concentrate on medical domain should be first constructed. In recent years, kinds of datasets have been proposed for the Medical VQA task. Most of these datasets focus on radiology images (including CTs, X-rays and MRIs) such as VQA-RAD, VQA-Med (2018, 2019, 2020) and SLAKE, while PathVQA dataset focuses on pathology images. Different from other datasets that only have triplets of image-question-answer, SLAKE is more comprehensive which has both semantic labels (e.g., labeled segmentation or bounding boxes of objects in the medical images) and structural medical knowledge base. Besides, SLAKE is also a bilingual dataset of both English and Chinese.

Table 11.1 Major datasets for medical VQA and their main characteristics

Dataset	Source of images	Number of images	Number of questions	Evaluation metrics
VQA-Med-2018 [8]	PubMed Central articles	2,866	6,413	BLEU & WBSS & CBSS
VQA-Med-2019 [3]	MedPix database	4,200	15,292	BLEU & Acc.
VQA-Med-2020 [1]	MedPix5 database	5,000	5,000	BLEU & Acc.
VQA-RAD [13]	MedPix database	315	3,515	BLEU & Acc.
PathVQA [9]	Textbook of Pathology & Basic Pathology	1670	32,799	BLEU & Acc.
SLAKE [14]	[11, 20, 24]	642	14,028	BLEU & Acc.

Details of these six datasets are presented in the following text, and the main characteristics of which are summarized in Table 11.1.

VQA-Med-2018

VQA-Med-2018 [8] is the first medical VQA dataset. It contains 6,413 questions-answer pairs, and 2,866 medical images extracted from PubMed Central articles. The question-answer pairs are generated from captions of the medical images via a semi-automatic approach. First, all possible question-answer pairs from captions are generated using a rule-based question generation (QG) system. The systems contain four modules: sentence simplification, answer phrase identification, question generation and candidate questions. Then, since the candidate questions generated by automatic approach may be noisy because the rules defined may not adequately capture the complex characteristics of medical field terminologies, two expert human annotators manually checked all the generated question-answer pairs associated with the medical images in two passes. In the first pass, one annotator proofreads all the question answer pairs to ensure grammatical and semantic correctness. In the second pass, another annotator verifies that all the question-answers are correct, ensuring that they are clinically relevant to the medical images. The overall set is split into 5,413 question answer pairs (associated with 2,278 medical images) for training, 500 question answer pairs (associated with 324 medical images) for validation and 500 questions (associated with 264 medical images) for testing.

VQA-Med-2019

VQA-Med-2019 [3] consists of 4,200 radiology images and 15,292 question-answer pairs. This dataset focuses on four categories of clinical questions: modality, plane, organ system and abnormality. These categories have different levels of difficulty and utilize classification and text generation methods. And all questions can be answered from the image content without requiring any additional medical knowledge or domain specific reasoning. The training set contains 3,200 images and 12,792 question-answer (QA) pairs, with three to four questions per image. The validation set contains 500 medical images and 2,000 QA pairs. The test set consists of 500 medical images and 500 questions.

VQA-Med-2020

For VQA-Med-2020 dataset [1], the images are selected from MedPix5 database for image-based diagnosis of relevant medical images. Diagnostic methods selected include CT/MRI imaging, angiography, characteristic imaging appearance, radiographs, imaging features, ultrasound and diagnostic radiology. Each problem appears in the created VQA data at least 10 times. The training set contains 4,000 radiology images and 4,000 question-answer pairs. The validation set consists of 500 radiology images and 500 question answer pairs. The test set consists of 500 radiology images and 500 questions.

VQA-RAD

VQA-RAD [13] contains 3.5K clinician annotated question-answer pairs and 315 images from MedPix31. It is the first manually constructed dataset in which clinicians asked naturally occurring questions regarding radiology images and provided reference answers. Questions can be classified into modality, plane, organ system, abnormality, object/condition presence, positional reasoning, color, size, attribute other, counting and others.

PathVQA

PathVQA [9] consists of 32,799 question-answer pairs generated from 1,670 pathology images collected from two pathology textbooks: "Textbook of Pathology" and "Basic Pathology", and 3,328 pathology images collected from the PIER7 digital library. There are seven categories of questions: what, where, when, whose, how, how much/how many, and yes/no. The first six categories pertain to 16,465 open-ended questions, accounting for 50.2% of all questions. The remaining questions are close-ended "yes/no" questions. The numbers of "yes" and "no" answers are balanced as 8,145 and 8,189, respectively.

SLAKE

SLAKE [14] contains 642 images with 14,028 question-answer pairs and 5,232 medical knowledge triplets for the training and evaluation of medical VQA models. Question generation uses an annotation system. The system starts with a pre-defined question template for each body part. Each template provides a number of candidate questions for each content type. SLAKE has ten content types (e.g., modality, position and color) for the questions. The images are split into 450 for training, 96 for validation and 96 for testing.

11.3 Classical VQA Methods for Medical VQA

In this section, we will review classical VQA methods that is widely used in general domain and quickly adapted in Medical VQA tasks. These methods usually use convolution networks (CNNs) such as VGGNet and ResNet networks to extract visual features from medical images, recurrent networks (RNNs) such as LSTM, Bi-LSTM or GRU networks to extract textual features from clinical-related questions, classical multi-modal feature fusion strategies such as joint embedding and attention mechanisms, and a multi-layer classifier or sequence-to-sequence encoder-decoder to predict the answer.

Motivation

Inspired by the success application of VQA in general domain, Medical VQA is proposed to help relieve the pressure of healthcare systems. Medical VQA task can provide a "second opinion" to radiologists regarding their analysis of the image, and the responses can be used by patients to obtain basic information regarding the medical image, without consulting their doctor. As a domain-specific branch of general VQA task, some classical VQA methods in general domain can be quickly adapted to solve Medical VQA tasks.

Methods

As shown in Fig. 11.1, classical VQA methods usually consist of four primary components: a pre-trained CNN-based image feature extractor, a RNN-based question feature extractor, a classical multi-modal feature fusion module and a classifier or a generator to predict the answer.

For multi-modal feature fusion, **joint embedding** is a simple but effective approach, including element-wise operation, concatenation and bilinear pooling.

Thanki et al. [22] proposed an encoder-decoder architecture, where pre-trained CNN networks on ImageNet such as VGG19 and DenseNet-201 are used to extract visual features from the medical images, and pre-trained word embedding on PubMed articles along with a 2-layer LSTM network are used to extract textual features from the question. For multi-modal features fusion, a simple element-wise multiplication is applied. Finally, the fused features are passed into a LSTM decoder to generate language in natural language.

Allaouzi et al. [4] used a pre-trained VGG16 network to extract visual features from the medical images, word embedding followed with a Bi-directional LSTM

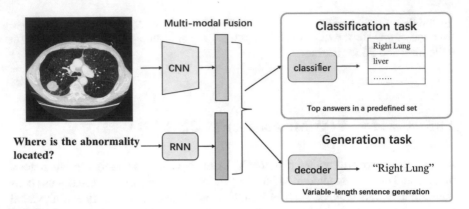

Fig. 11.1 Framework of classical VQA methods adopted for Medical VQA

to embed words in the corresponding questions and extract textual features from questions. Then, the features of both medical images and questions are concatenated and passed through dense layers to get a length-fixed vector as the multi-modal feature vector. At last, a Decision Tree Classifier is used to predict answers based on this multi-modal feature vector.

Afterward, Allaouzi et al. [5] proposed an encoder-decoder model that predicts answers by generating each word of the answer. A pre-trained DenseNet-121 network and pre-trained word embedding are used to extract image and question features respectively. These multi-modal features are concatenated as QI vector, which will be then concatenated with features of generated words as encoder vector. A decoder is used to generate answers according to the encoder vector. Similarly, Talafha et al. [21] proposed an encoder-decoder sequence-to-sequence architecture, in which pre-trained VGGNet is utilized instead of DesNet.

Abacha et al. [2] adopted the multimodal compact bilinear pooling (MCB) strategy to fuse multi-modal features. ResNet and LSTM are used to extract image and question features, and these single-modal features are combined by MCB as multi-modal feature. For answer prediction, a classifier is used.

Attention mechanisms are more complicated and powerful in fusing multi-modal features than joint embedding methods.

Zhou et al. [27] employed a basic attention module to fuse image and question features, enhancing the model's capability of learning and generalization. Specifically, the Inception-ResNet-v2 network extracts visual features from medical images and a Bi-directional LSTM network extract features from questions. The attended visual features are concatenated with question features as multi-modal feature, which is then passed through a classifier to predict the answer.

Abacha et al. [2] used stacked attention network (SAN) to implement multi-step reasoning in Medical VQA tasks. VGG16 and LSTM are used to extract single-modal features respectively. Two attention layers are utilized in the adopted SAN to fuse image and question features. For answer prediction, a one-layer neural network with softmax is used as a classifier.

Peng et al. [17] proposed a framework that employed co-attention mechanism with bilinear pooling. ResNet-152 pre-trained on ImageNet is used to extract visual features, and the word embedding pre-trained on medical-relevant copora is used followed by a LSTM network to extract question features. Then, a co-attention is implemented on the image and question features, where attended image features and attended question features are obtained. To further fuse the attended multi-modal features, Multi-modal Factorized Bilinear Pooling (MFB) is used.

Furthermore, Shi et al. [19] used Multi-modal Factorized High-order pooling (MFH) instead of MFB for multi-modal feature fusion. Besides, more information sources are taken into account for feature extraction, including question category and question topic distribution. To be specific, the attended image features and the attended question features obtained by co-attention mechanism are fused with the above two extra features using MFH. The fused multi-modal features are used to predict the answer.

Performance and Limitations

It is simple and efficient to adapt classical VQA methods for Medical VQA tasks. Even the joint embedding method could make a competitive baseline for Medical VQA. Similar to the findings in general VQA, the more comprehensive and effective the fusion methods are, the better performance the model will achieve. The bilinear pooling methods perform better than the simple element-wise operation and concatenation, attention-based methods combined with bilinear pooling performs better than single attention mechanism or single bilinear pooling methods. However, most of these methods use pre-trained CNNs on ImageNet to extract visual features from medical images, which will have a big gap between the general images and medical images. Thus, the performance may be affected during this transfer learning process.

11.4   Meta-Learning Methods for Medical VQA

In this section, we will review the meta-learning methods utilized for Medical VQA tasks. Specifically, we give the detailed description of the first proposed framework based on meta-learning, namely Mixture of Enhanced Visual Features (MEVF). Then, we introduce a variant of MEVF with conditional reasoning and the other more advanced meta-learning method called Multiple Meta-model Quantifying for Medical VQA (MMQ).

Motivation

Large amount of labeled data are often required to train VQA models in general domain. However, in medical domain, it is not as easy building large-scale and

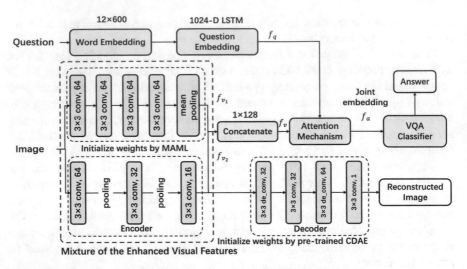

Fig. 11.2 Overview of the meta-learning method of MEVF for Medical VQA

well-elaborated Medical VQA datasets as building large-scale general VQA datasets. In other words, large-scale labeled data are usually lacked for Medical VQA tasks. Thus, to overcome the limitation of data shortage, meta-learning methods are adopted in solving Medical VQA tasks.

Methods

Nguyen et al. [16] first adopted meta-learning in Medical VQA tasks. As shown in Fig. 11.2, the core of the proposed Medical VQA framework is Mixture of Enhanced Visual Feature (MEVF), of which the weights of are initialized by Model-Agnostic Meta-Learning (MAML) and Convolutional Denoising Auto-Encoder (CDAE) pre-trained on medical images. Then, it will be fine-tuned on Medical VQA datasets in an end-to-end way. In fact, when adapting MAML to a new task, the meta-parameters of MAML can be quickly adapted to the new task. To train MAML, images in VQA-RAD dataset are manually reviewed and classified into 9 categories. During each training iteration of MAML, 5 tasks are sampled and for each task 3 classes are randomly sampled. For each class, 6 images are randomly sampled to update models. To train CDAE, more than 10K radiology images are collected from the Internet. Specifically, reconstruction error between the original and reconstructed radiology images is minimizing when training CDAE. The objective of overall framework is a multi-task loss function, which takes into account both VQA classification loss and the reconstruction loss.

Based on the above Medical VQA framework, Zhan et al. [26] proposed a novel conditional reasoning method for medical VQA, aiming to automatically learn effective reasoning skills for various Medical VQA tasks. Particularly, a question-conditioned reasoning module is used to guide the importance selection over multimodal fusion features. Besides, considering the different natures of close-ended and open-ended Medical VQA tasks, an extra type-conditioned reasoning module is proposed to separately learn reasoning skills for different sets of the two types of Medical VQA tasks.

However, MAML suffers from meta-annotation phase for medical image dataset. More recently, Do et al. [7] presented a novel multiple meta-model quantifying method (MMQ) to enhance the metadata by automatically annotating, handle noisy labels during meta-agnostic process and select meta-models with robust features for downstream Medical VQA tasks. Different from MEVF, MMQ only utilizes MAML as primary component without CDAE and selects multiple meta-models rather than only one MAML model. While the meta-training process is similar to MEVF. Then, the trained meta-models are used to refine initial dataset by auto-annotating labels and dealing with noisy labels. At last, these meta-models are scored to select which is useful for Medical VQA tasks.

Performance and Limitations

The novel meta-learning methods could overcome the lack of large-scale labeled data in Medical VQA datasets. Compared to the classical VQA methods directly adapted in Medical VQA, these methods could make better use of medical images and learn more effectively from limited data, which will result in better performance. However, these methods did not make use of the prevalent and powerful Transformer and BERT, which may further improve the performance.

11.5   BERT-Based Methods for Medical VQA

In this section, we will review BERT-based methods proposed for Medical VQA. We first introduce methods that simply take BERT as language encoder for clinically-related questions, while share similar fusion architectures with other classical VQA methods. Then, we present detailed description of BERT-based methods that utilize Transformer layers to interact between two modalities of image and question. At last, we show methods that pre-train BERT-like models on medical image-text pairs and fine-tune the pre-trained models on Medical VQA datasets.

Motivation

Recently, a large amount of studies have demonstrated the successful application of BERT in Natural Language Processing (NLP) and Vision-Language tasks in general domain. It is with impressive effectiveness either to simply use BERT as language encoder to extract textual features or to use BERT to preform interactions between multi-modal features. Thus, it is natural to adopt the prevalent and powerful BERT to learn better representations of both medical images and clinical-relevant questions in Medical VQA tasks.

Methods

The most widely used way of BERT in Medical VQA tasks is simply taking BERT as language encoder to learn textual embeddings of clinical-related questions while sharing similar multi-modal fusion architecture with other classical VQA methods for Medical VQA tasks. Zhou et al. [28] proposed TUA1 which uses Inception-ResNet-V2 to extract visual features from the medical images and BERT to extract textural features from the corresponding question. In addition, TUA1 adopts the sub-task strategy, which first uses a simple classifier to identify the category of each question. For the question type of Abnormality, TUA1 uses a sequence-to-sequence generation model to predict the answer. While for other types of question, TUA1 uses a classifier to predict the answer. Vu et al. [23] proposed a method of ensembling models that either uses Skip-thought vectors or BERT to extract question features and utilizes multi-glimpse attention mechanisms with bilinear fusion. Yan et al. [25] proposed Hanlin that uses a modified VGG16 network with Global Average Pooling strategy to extract visual features, BERT to extract question features and MFB with co-attention to fuse multi-modal features. Instead of using BERT pre-trained on general domain to extract question features, Jung et al. [10] and Chen et al. [6] adopted the domain-specific BioBERT that is pre-trained on large-scale bio-medical corpora.

Different from the above methods, as shown in Fig. 11.3, Ren and Zhou [18] proposed a new classification and generative model for Medical VQA (CGMVQA) which utilizes a 4-layer Transformer to interact between multi-modal features. CGMVQA adopts the sub-task strategy which first uses a classifier to identify the type of each question. CGMVQA uses ResNet152 to extract visual features from medical images. Specifically, to obtain richer semantic information from different dimensions, the visual features are extracted from different convolution layers of ResNet152 network. Followed by Fully Convolutional Networks and Global Average Pooling, the output of each 5 blocks in ResNet152 network is taken as visual token that will be passed through Transformer layers. CGMVQA uses Word Pieces to tokenize questions and BERT to embed questions. For Abnormality questions,

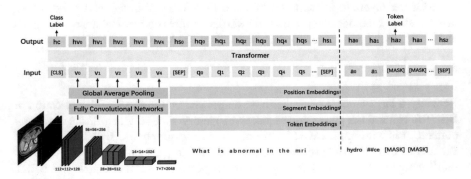

Fig. 11.3 Overview of the CGMVQA proposed for Medical VQA

CGMVQA acts as a generator, where answers are predicted based on the output features of masked tokens. For other types of questions, CGMVQA acts as a classifier, where answers are predicted based on the output features of special token [CLS].

Furthermore, Khare et al. [12] proposed multimodal BERT pre-training for Medical VQA, which pre-trains a model that shares the similar architecture with CGMVQA on medical image-caption pairs from ROCO dataset, and then fine-tunes the model on downstream VQA-Med-2019 and VQA-RAD datasets. During pre-training, Masked Language Modeling (MLM) task is utilized. Specifically, to better learn medical knowledge, only medical keywords are masked from captions in ROCO and the model only needs to predict these masked tokens. During fine-tuning, instead of using the output feature of [CLS] token, MMBERT uses the average pooling of the last year of the Transformer as fused multi-modal feature. Then, this multi-modal feature passes through dense layers for answer classification.

Similar to MMBERT, Moon et al. [15] proposed a vision-language pre-training for medical images and text called MedViLL. Different from MMBERT, MedViLL uses ResNet50 of which the last feature map ($16 \times 16 \times 2048$) is flattened as visual features of medical images. Besides, MedViLL is pre-trained on MIMIC-CXR and Open-I datasets. During pre-training, Masked Language Modeling (MLM) and Image Report Matching (IRM) tasks are used. During fine-tuning, the output feature of [CLS] token is used to predict the answer.

Performance and Limitations

So far, the BERT-based methods have achieved state-of-the-art performance than other kinds of methods. However, the visual feature extractors of these methods are still pre-trained CNNs on ImageNet such as ResNet, which may neglect the characteristics of medical images and may have a negative impact on the performance.

References

1. A.B. Abacha, V. Datla, S.A. Hasan, D. Demner-Fushman, H. Müller, Overview of the vqa-med task at imageclef 2020: visual question answering and generation in the medical domain, in *CLEF* (2020)
2. A.B. Abacha, S. Gayen, J. Lau, S. Rajaraman, D. Demner-Fushman, Nlm at imageclef 2018 visual question answering in the medical domain, in *CLEF* (2018)
3. A.B. Abacha, S.A. Hasan, V. Datla, J. Liu, D. Demner-Fushman, H. Müller, Vqa-med: overview of the medical visual question answering task at imageclef 2019, in *CLEF* (2019)
4. I. Allaouzi, M. Ahmed, Deep neural networks and decision tree classifier for visual question answering in the medical domain, in *CLEF* (2018)
5. I. Allaouzi, M. Ahmed, B. Benamrou, An encoder-decoder model for visual question answering in the medical domain, in *CLEF* (2019)
6. G. Chen, H. Gong, G. Li, Hcp-mic at vqa-med 2020: effective visual representation for medical visual question answering, in *CLEF* (2020)
7. T. Do, B.X. Nguyen, E. Tjiputra, M.-N. Tran, Q.D. Tran, A. Nguyen, Multiple meta-model quantifying for medical visual question answering, arXiv:2105.08913 (2021)

8. S.A. Hasan, Y. Ling, O. Farri, J. Liu, H. Müller, M. Lungren, Overview of imageclef 2018 medical domain visual question answering task, in *CLEF* (2018)
9. X. He, Y. Zhang, L. Mou, E. Xing, P. Xie, Pathvqa: 30000+ questions for medical visual question answering, arXiv:2003.10286 (2020)
10. B. Jung, L. Gu, T. Harada, bumjun_jung at vqa-med 2020: Vqa model based on feature extraction and multi-modal feature fusion, in *CLEF (Working Notes)* (2020)
11. A.E. Kavur, N. Gezer, M. Baris, P.-H. Conze, V. Groza, D.D. Pham, S. Chatterjee, P. Ernst, S. Özkan, B. Baydar, D. Lachinov, S. Han, J. Pauli, F. Isensee, M. Perkonigg, R. Sathish, R. Rajan, S. Aslan, D. Sheet, G. Dovletov, O. Speck, A. Nürnberger, K. Maier-Hein, G. Akar, G.B. Ünal, O. Dicle, M.A. Selver, Chaos challenge - combined (ct-mr) healthy abdominal organ segmentation. Med. Image Anal. **69**, 101950 (2021)
12. Y. Khare, V. Bagal, M. Mathew, A. Devi, U.D. Priyakumar, C.V. Jawahar, Mmbert: multimodal bert pretraining for improved medical vqa, in *2021 IEEE 18th International Symposium on Biomedical Imaging (ISBI)* (2021), pp. 1033–1036
13. J. Lau, S. Gayen, A.B. Abacha, D. Demner-Fushman, A dataset of clinically generated visual questions and answers about radiology images. Sci. Data **5** (2018)
14. B. Liu, L.-M. Zhan, L. Xu, L. Ma, Y. Yang, X.-M. Wu, Slake: a semantically-labeled knowledge-enhanced dataset for medical visual question answering, in *2021 IEEE 18th International Symposium on Biomedical Imaging (ISBI)* (2021), pp. 1650–1654
15. J.H. Moon, H. Lee, W. Shin, E. Choi, Multi-modal understanding and generation for medical images and text via vision-language pre-training. *CoRR* (2021)
16. B.D. Nguyen, T. Do, B.X. Nguyen, T. Do, E. Tjiputra, Q.D. Tran, Overcoming data limitation in medical visual question answering, in *MICCAI* (2019)
17. Y. Peng, F. Liu, Umass at imageclef medical visual question answering(med-vqa) 2018 task, in *CLEF* (2018)
18. F. Ren, Y. Zhou, Cgmvqa: a new classification and generative model for medical visual question answering. IEEE Access **8**, 50626–50636 (2020)
19. L. Shi, F. Liu, M. Rosen, Deep multimodal learning for medical visual question answering, in *CLEF* (2019)
20. A. Simpson, M. Antonelli, S. Bakas, M. Bilello, K. Farahani, B. Ginneken, A. Kopp-Schneider, B. Landman, G. Litjens, B. Menze, O. Ronneberger, R. Summers, P. Bilic, P. Christ, R. Do, M. Gollub, J. Golia-Pernicka, S. Heckers, W. Jarnagin, M. McHugo, S. Napel, E. Vorontsov, L. Maier-Hein, M.J. Cardoso, A large annotated medical image dataset for the development and evaluation of segmentation algorithms. arXiv:1902.09063 (2019)
21. B. Talafha, M. Al-Ayyoub, Just at vqa-med: a vgg-seq2seq model, in *CLEF* (2018)
22. A. Thanki, K. Makkithaya, Mit manipal at imageclef 2019 visual question answering in medical domain, in *CLEF* (2019)
23. M.H. Vu, R. Sznitman, T. Nyholm, T. Löfstedt, Ensemble of streamlined bilinear visual question answering models for the imageclef 2019 challenge in the medical domain, in *CLEF* (2019)
24. X. Wang, Y. Peng, L. Lu, Z. Lu, M. Bagheri, R. Summers, Chestx-ray8: hospital-scale chest x-ray database and benchmarks on weakly-supervised classification and localization of common thorax diseases, in *2017 IEEE Conference on Computer Vision and Pattern Recognition (CVPR)* (2017), pp. 3462–3471
25. X. Yan, L. Li, C. Xie, J. Xiao, L. Gu, Zhejiang university at imageclef 2019 visual question answering in the medical domain, in *CLEF* (2019)
26. L.-M. Zhan, B. Liu, L. Fan, J. Chen, X.-M. Wu, Medical visual question answering via conditional reasoning, in *Proceedings of the 28th ACM International Conference on Multimedia* (2020)
27. Y. Zhou, X. Kang, F. Ren, Employing inception-resnet-v2 and bi-lstm for medical domain visual question answering, in *CLEF* (2018)
28. Y. Zhou, X. Kang, F. Ren, Tua1 at imageclef 2019 vqa-med: a classification and generation model based on transfer learning, in *CLEF* (2019)

Chapter 12
Text-Based VQA

Abstract VQA requires reasoning regarding the visual content of an image. However, in a large proportion of images, visual content is not the only information. Texts that can be recognized by optical character recognition (OCR) tools provide considerably more useful and high-level semantic information, such as the street name, product brand and prices, which is not available in any other forms in the scene. Interpreting this written information in human environments is essential for performing most everyday tasks like making a purchase, using public transportation and finding a location in a city. Hence, the new task TextVQA has been proposed. In this chapter, we briefly introduce the main datasets that benchmark progress in this field, including TextVQA [29], ST-VQA [2] and OCR-VQA [25]. Subsequently, we describe an important tool (OCR) that is a prerequisite for the reasoning process, as texts must be first recognized. Next, we select 3 representative and effective models to address this problem and describe them in a sequential manner.

12.1 Introduction

One benefit of visual question answering is that it can help visually impaired users be aware about their surroundings. As shown in the VizWiz study [9], up to 21% of these questions involve reading and analyzing the text captured in a user's surroundings: 'what temperature is my oven set to?' or 'what denomination is this bill?' To answer these questions, the model must have the following abilities:

- to realize when the question includes text,
- to detect image regions containing text,
- to convert pixel representations (convolutional features) of these regions to symbols or textual representations (semantic word embeddings),
- to jointly analyze the detected text and visual content,
- to decide if the detected text needs to be 'copy-pasted' as the answer or if the detected text informs the model about an answer in the answer space.

Notably, the existing VQA models cannot answer such questions, as all of the above skills cannot simply be integrated into a monolithic network. To address this new challenge, several datasets have been proposed to evaluate the overall performance in

Q. Wu et al., *Visual Question Answering*, Advances in Computer Vision and Pattern Recognition, https://doi.org/10.1007/978-981-19-0964-1_12

terms of the above mentioned abilities, along with several classic baseline models. To accomplish reasoning, the model must be able to read. Optical character recognition (OCR) is a subfield of computer vision and involves many mature algorithms. For a model to be able to read is a simple task, as only an independent OCR module needs to be added. We discuss the implementation and importance of the module in the subsequent sections. Moreover, we highlight the main methods that have been introduced in the field of TextVQA and can be categorized as simple fusion models, transformer-based models and graph-based models.

12.2   Datasets

Questions that require reading and reasoning are uncommon in the standard VQA datasets, as they are not collected in a setting similar to those of visually impaired users. The existing relevant dataset VizWiz [9] is small in size, which renders its application as a benchmark challenging. To focus on the understanding and reasoning of scene texts in images, several datasets have been proposed. We review the existing TextVQA datasets, describe how the datasets are created and perform a comparative analysis. A coarse comparison of different datasets is presented in Table 12.1.

12.2.1   TextVQA

To study the task of answering questions that require reading text in images, a new dataset has been established, which is publicly available at the TextVQA official website.[1]

The TextVQA dataset contains 28,408 images from the Open Images dataset [18] (from categories that tend to contain text e.g."billboard", "traffic sign", and "white board"), with each question accompanied by 10 human annotated answers. The final accuracy is calculated through soft voting of the 10 answers, similar to VQAv2 [8].

The training and validation set is collected from the training set of the Open Images' training set, and the test set is collected from the Open Images' test set. To automatically select appropriate images from this large source, an OCR model

Table 12.1 Comparison of three TextVQA datasets

Dataset	Images	Questions
TextVQA	28,408	45,336
ST-VQA	23,038	31,791
OCR-VQA	207,572	1,002,146

[1] https://textvqa.org.

Rosetta [3] is applied to these images, which computes the average number of OCR boxes in each category. The categories with the most OCR boxes are selected.

The dataset construction involved three processes. In the first stage, images that did not contain text were filtered by human annotators. In the second stage, 1–2 questions were provided by annotators. In the third stage, 10 answers for each question were collected from the annotators, similar to the VQA [1, 8] and VizWiz [9] dataset settings. The quality of data was ensured by additional filtering performed by annotators. In addition, handcrafted questions were considered, and correct answers were expected to filter the inferior annotators.

In addition, a model known as LoRRA was proposed to be used as the baseline of this new dataset, as described in Sect. 12.4.

12.2.2 ST-VQA

To highlight the importance of high-level semantic information in the VQA process, a new dataset, *scene text visual question answering* (ST-VQA), was proposed. In this dataset, questions could only be answered based on the text present in the image.

In particular, the ST-VQA dataset [2] contains natural images from multiple sources, including ICDAR 2013 [17], ICDAR2015 [16], ImageNet [5], VizWiz [9], IIIT STR [24], Visual Genome [19] and COCO-Text [30]. The questions and answers were collected through Amazon Mechanical Turk. The format of the ST-VQA is similar to that of the TextVQA dataset. Instead of 10 human annotated answers for the TextVQA dataset, in ST-VQA, each question is accompanied by only one or two ground-truth answers provided by the question writer. The ST-VQA dataset involves three tasks, which gradually increase in difficulty: Task 1, strongly contextualized provides a dynamic candidate dictionary of 100 words per image, while Task 2, weakly contextualized provides a fixed answer dictionary of 30,000 words for the whole dataset. For Task 3, open dictionary, the model is supposed to generate an answer without additional information.

The ST-VQA dataset consists of 23,000 images with up to three questions/answer pairs per image. A train and test split is implemented. The training set consists of 19,000 images with 26,000 questions, while the test set consists of 3,000 images with 4,000 questions per task.

To automatically select images from these sources, ST-VQA uses an end-to-end single shot text retrieval architecture to choose images containing at least 2 text instances.

To account for reasoning errors and text recognition errors, a new evaluation metric, average normalized Levenshtein similarity (ANLS), is adopted by ST-VQA as the official evaluation metric. The metric is defined as scores $1 - d_L(a_{pred}, a_{gt})/max(|a_{pred}|, |a_{gt}|)$ (where a_{pred} and a_{gt} denote the prediction and ground-truth answers, respectively, and d_L is the edit distance) averaged over all questions. Additionally, all scores below the threshold of 0.5 are truncated to 0 before averaging.

The dataset, performance evaluation scripts and an online evaluation service are available through the ST-VQA Web portal.[2]

12.2.3   OCR-VQA

The OCR-VQA dataset [25] contains 207,572 images of book covers, with template-based questions querying the title, author, edition, genre, year or other information regarding the book. Each question has a single ground-truth answer, and the dataset assumes that the answers to these questions can be inferred from the book cover images.

This dataset can be explored and downloaded from the project website.[3]

12.3   OCR Token Representation

Optical character recognition (OCR) is a mature subfield of computer vision, aimed at detecting and recognizing text. This task sets the starting point for a generalized VQA system that can integrate the reading ability.

In the field of OCR, the commonly used methods can be divided into two parts: text detection and recognition. To address the problem of text detection, several methods [11, 20, 21, 32] that are based on fully convolutional neural networks have been proposed. Text recognition methods such as [14] address text recognition as a classification problem in a word-by-word manner. Connectionist temporal classification (CTC) has also been widely used in scene text recognition [3, 22, 27, 31]. More recent methods [4, 10, 23] focus on the end-to-end architecture, which mostly consists of a convolutional neural network (CNN) as an encoder and a long short-term memory (LSTM) as a decoder.

In typical TextVQA models, images are first processed by an independent OCR model to yield OCR tokens, which are then encoded into OCR token representations. Intuitively, to represent text in images, it is necessary to encode not only the text characters but also the text appearance (e.g.color, font, and background) and its spatial location in the image (e.g.words appearing on the top of a book cover are more likely to be book titles).

After obtaining a set of N OCR tokens in an image through external OCR systems, from the nth token (where $n = 1, , N$), multiple OCR token representations can be extracted:

FastText feature. Generated by using pretrained FastText embeddings [15], which can produce word embeddings even for out-of-vocabulary (OOV) tokens. This word embedding is a 300-dimensional vector that contains subword information.

---

[2] https://rrc.cvc.uab.es/?ch=11.

[3] https://ocr-vqa.github.io/.

Faster R-CNN feature. Appearance feature obtained from the Faster R-CNN detector, extracted via RoI-Pooling on the OCR token's bounding box, which has 2048 dimensions.

Pyramidal histogram of characters (PHOC) feature. A 604-dimensional vector that captures the characters that are present in the token. This feature is more robust to OCR recognition errors and can be considered a coarse character model.

Location feature. This 4-dimensional location feature is based on the OCR token's relative bounding box coordinates $[x_{min}/W_{im}, y_{min}/H_{im}, x_{max}/W_{im}, y_{max}/H_{im}]$, where W_{im} and H_{im} denote the image width and height, respectively.

The use of rich representations for OCR tokens was first proposed by M4C [13], which corresponded to a considerable enhancement in performance.

12.4   Simple Fusion Models

The simplest and most direct method is based on the simple pairwise fusion of two modalities. In this section, we introduce the first simple fusion model named LoRRA, which has been proposed as the baseline model for the TextVQA [29] dataset.

12.4.1   LoRRA: Look, Read, Reason & Answer

Look, read, reason & answer (LoRRA) [29] was proposed at the same time as the TextVQA dataset to be set as a baseline. The code was originally published in the Pythia framework and subsequently integrated into the more general mmf [28] framework.[4]

At a high level, LoRRA contains three components: (i) a VQA component to analyze and infer the answer based on the image v and question q, (ii) a reading component that allows the model to read the text in the image, and (iii) an answering module that obtains predictions from an answer space or points to the text read by the reading component. The overall model is shown in Fig. 12.1. Thus, LoRRA is a simple extension of previous VQA models, with an additional OCR attention branch. Consequently, this model adds OCR tokens as a dynamic vocabulary to the answer classifier and uses a copy mechanism to choose a single OCR token.

VQA Component The question words of question q are first embedded with a predefined embedding function GloVe [26] and then encoded iteratively with a long short-term memory (LSTM [12]) recurrent network to produce a question embedding $f_Q(q)$. For images, two kinds of representations exist for visual features: grid-based convolutional features and/or features extracted from the bounding box proposals. These features are referred to as $f_I(v)$, where f_I is the network that extracts the image representation. A simple attention mechanism f_A is used to assign a weighted

---

[4] https://github.com/facebookresearch/mmf.

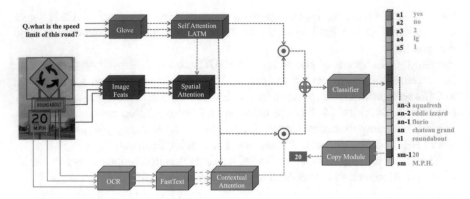

Fig. 12.1 Overview of the look, read, reason & answer (LoRRA) approach. The approach looks at the image, reads its text, analyzes the image and text content and answers, either with an answer a from the fixed answer vocabulary or by selecting one of the OCR strings s. Dashed lines indicate components that are not jointly trained. The answer cubes on the right with darker color correspond to a higher attention weight. The OCR token '20' has the highest attention weight in the example

average over the visual features based on $f_I(v)$ and $f_Q(q)$ as the output. The output is combined with question embedding to calculate the VQA features $f_{VQA}(v, q)$.

$$f_{VQA}(v, q) = f_{comb}(f_A(f_I(v), f_Q(q)), f_Q(q)) \tag{12.1}$$

A feedforward MLP is applied to the combined embedding to predict the possibility of a certain answer being correct.

Reading Component To allow a model to read text from an image, an independent OCR model that is not jointly trained with the whole system is used. The assumption is that the OCR model can read and return word tokens from an image.

In the process of weighted attention, because the features are multiplied by weights and averaged, the ordering information is lost. To provide the answer module with the ordering information of the original OCR tokens, the attention weights are concatenated with the final weight-averaged features. This framework allows the answer module to identify the original attention weights for each token in order.

Answer Module With a ?xed answer space, the existing VQA models can only predict fixed tokens, which limits their generalization to out-of-vocabulary (OOV) words. As the text in images frequently contains words not encountered during training, it is difficult to answer text-based questions based on only a prede?ned answer space. To ensure generalization to arbitrary text, the authors take inspiration from pointer networks that allow for pointing to OOV words in context. The authors extend the answer space through the addition of a dynamic component that corresponds to M OCR tokens. The model can thus predict probabilities $(p_1, \ldots, p_N, \ldots, p_{N+M})$ for $N + M$ items in the answer space instead of the original N items.

12.5   Transformer-Based Models

The transformer architecture has been widely used since it was first proposed. The architecture exhibits a satisfactory performance across many multimodal tasks, with no exception to TextVQA. In this section, we introduce the first model, named the multimodal multicopy mesh (M4C) [13]. This model exploits the popular transformer architecture to fuse and jointly model multiple modalities and iteratively generate answers, which also enable the model to generate answers with more than one word. This model exhibits high performance on various datasets.

12.5.1   Multimodal Multicopy Mesh Model

The M4C model involves three enhancements. The first enhancement pertains to the adoption of the transformer architecture, which allows for the homogeneous natural fusion of different modalities. The second enhancement pertains to the introduction of and emphasis on many rich representations of OCR tokens, as described in Sect. 12.3. These additional OCR token representations promote the exploitation of OCR token information from several aspects. The third enhancement pertains to the use of the iterative decoder along with a dynamic pointer network for answer decoding. The overall model of M4C is shown in Fig. 12.2.

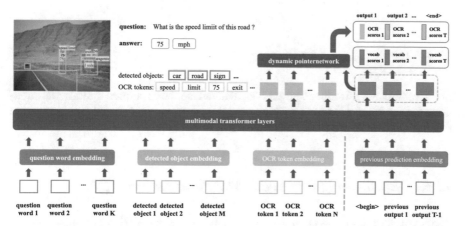

Fig. 12.2 Overview of the M4C model. All entities (question words, detected visual objects, and detected OCR tokens) are projected into a common d-dimensional semantic space through domain-specific embedding approaches, and multiple transformer layers are applied over the list of projected entities. Based on the transformer outputs, the answer is predicted through iterative autoregressive decoding, where at each step, the model either selects an OCR token through the dynamic pointer network or a word from its fixed answer vocabulary

Instead of custom pairwise fusion mechanisms between a pair of two modalities, the multimodal multicopy mesh (M4C) model naturally fuses different modalities homogeneously by using a multimodal transformer architecture.

Moreover, M4C does not simply view TextVQA as a simple classification problem, which is restricted to one prediction step. Instead, the model enables iterative answer decoding with a dynamic pointer network owing to the natural decoding ability of the transformer architecture. The dynamic pointer network is implemented by calculating the dot product value (bilinear interaction) of the decoding output and output representation of each OCR token.

The question words are embedded into the corresponding sequence of d-dimensional feature vectors $\{x_k^{ques}\}$ (where $k = 1, , K$) using a pretrained BERT model. During training, the BERT parameters are fine-tuned using the question answering loss.

For visual objects, the set of M objects is obtained by a pretrained Faster R-CNN detector. The appearance feature x_m^{fr} is extracted using the detector's output from the mth object. The location feature x_m^b is defined as the relative bounding box coordinates of the mth object: $[x_{min}/W_{im}, y_{min}/H_{im}, x_{max}/W_{im}, y_{max}/H_{im}]$, where W_{im} and H_{im} are the image width and height, respectively.

For OCR tokens, M4C uses the rich representations, as described in Sect. 12.3.

All the entities (question words, visual objects, and OCR tokens) are embedded into the d-dimensional joint embedding space. Subsequently, a stack of L transformer layers is applied over these entities. Through the multihead self-attention mechanism in transformers, each entity can freely attend to all other entities, regardless of whether or not they pertain to the same entity.

12.6   Graph-Based Models

In this section, we introduce a graph-based model named structured multimodal attentions for TextVQA (SMA) [7], which uses the question-conditioned graph attention mechanism to enhance the textual-visual reasoning ability.

12.6.1   Structured Multimodal Attentions for TextVQA

The SMA first uses a structural graph representation to encode the object-object, object-text and text-text relationships appearing in the image and subsequently designs a multimodal graph attention network to perform the analysis. Finally, outputs from the above mentioned modules are processed using a global-local attentional answering module to produce an answer by iteratively splicing together tokens from both the OCR and general vocabulary.

At a high level, the SMA is composed of three modules: (1) a question self-attention module that decomposes questions into six subcomponents that have

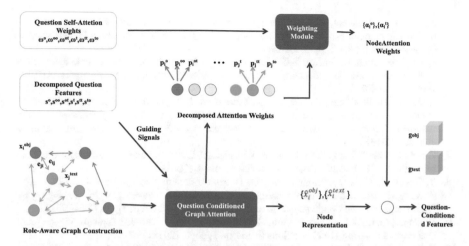

Fig. 12.3 Overview of the question-conditioned graph attention module. This module builds a heterogeneous graph whose mixed nodes are shown in different colors. Guiding signals help produce attention weights, which are fused with node representations to obtain question-conditioned features

different roles in the constructed object-text graph, (2) a question-conditioned graph attention module that reasons over the graph under the guidance of the question representations and infers the importance of different nodes as well as their relationships, and (3) a global-local attentional answering module that can generate answers with multiple words stitched together. The answering module of SMA is based on the iterative answer prediction mechanism in M4C with a modification of the first step input.

The key component of SMA, namely, the question conditioned graph attention module, is illustrated in Fig. 12.3.

Formally, given a question Q with T words $q = \{q_t\}_{t=1}^T$, $\{x_t^{bert}\}_{t=1}^T$ is obtained by using pretrained BERT [6]. The decomposed question features ($\mathbf{s}^o$, $\mathbf{s}^{oo}$, $\mathbf{s}^{ot}$, $\mathbf{s}^t$, $\mathbf{s}^{tt}$, $\mathbf{s}^{to}$) are considered question representations decomposed w.r.t object nodes (**o**), object-object (**oo**) edges, object-text (**ot**) edges, text nodes (**t**), text-text (**tt**) edges and text-object (**to**) edges.

References

1. S. Antol, A. Agrawal, J. Lu, M. Mitchell, D. Batra, C. Lawrence Zitnick, D. Parikh, VQA: visual question answering (2015), pp. 2425–2433
2. A.F. Biten, R. Tito, A. Mafla, L. Gomez, M. Rusinol, E. Valveny, C. Jawahar, D. Karatzas, Scene text visual question answering (2019)
3. F. Borisyuk, A. Gordo, V. Sivakumar, Rosetta: large scale system for text detection and recognition in images, in *Proceedings of the 24th ACM SIGKDD International Conference on Knowledge Discovery & Data Mining* (ACM, 2018), pp. 71–79

4. M. Busta, L. Neumann, J. Matas, Deep textspotter: an end-to-end trainable scene text local-ization and recognition framework, in *Proceedings of the IEEE International Conference on Computer Vision* (2017), pp. 2204–2212

5. J. Deng, W. Dong, R. Socher, L.-J. Li, K. Li, L. Fei-Fei, Imagenet: a large-scale hierarchical image database, in *2009 IEEE Conference on Computer Vision and Pattern Recognition* (IEEE, 2009), pp. 248–255

6. J. Devlin, M.-W. Chang, K. Lee, and K. Toutanova, Bert: pre-training of deep bidirectional transformers for language understanding, in *Proceedings of the 2019 Conference of the North American Chapter of the Association for Computational Linguistics: Human Language Technologies* (2019)

7. C. Gao, Q. Zhu, P. Wang, H. Li, Y. Liu, A.V.D. Hengel, Q. Wu, Structured multimodal attentions for textVQA. IEEE Trans. Pattern Anal. Mach. Intell. (2021)

8. Y. Goyal, T. Khot, D. Summers-Stay, D. Batra, D. Parikh, Making the V in VQA matter: elevating the role of image understanding in visual question answering, in *Proceedings of the IEEE Conference on Computer Vision and Pattern Recognition* (2017), pp. 6904–6913

9. D. Gurari, Q. Li, A.J. Stangl, A. Guo, C. Lin, K. Grauman, J. Luo, J.P. Bigham, Vizwiz grand challenge: answering visual questions from blind people (2018)

10. T. He, Z. Tian, W. Huang, C. Shen, Y. Qiao, C. Sun, An end-to-end textspotter with explicit alignment and attention, in *Proceedings of the IEEE Conference on Computer Vision and Pattern Recognition* (2018), pp. 5020–5029

11. W. He, X.-Y. Zhang, F. Yin, C.-L. Liu, Deep direct regression for multi-oriented scene text detection, in *Proceedings of the IEEE International Conference on Computer Vision* (2017), pp. 745–753

12. S. Hochreiter, J. Schmidhuber, Long short-term memory. Neural Comput. **9**(8), 1735–1780 (1997)

13. R. Hu, A. Singh, T. Darrell, M. Rohrbach, Iterative answer prediction with pointer-augmented multimodal transformers for textVQA, in *Proceedings of the IEEE/CVF Conference on Computer Vision and Pattern Recognition* (2020), pp. 9992–10002

14. M. Jaderberg, K. Simonyan, A. Vedaldi, A. Zisserman, Reading text in the wild with convolutional neural networks. Int. J. Comput. Vis. **116**(1), 1–20 (2016)

15. A. Joulin, E. Grave, P. Bojanowski, T. Mikolov, Bag of tricks for efficient text classification (2017)

16. D. Karatzas, L. Gomez-Bigorda, A. Nicolaou, S. Ghosh, A. Bagdanov, M. Iwamura, J. Matas, L. Neumann, V.R. Chandrasekhar, S. Lu, et al., ICDAR 2015 competition on robust reading, in *2015 13th International Conference on Document Analysis and Recognition (ICDAR)* (IEEE, 2015), pp. 1156–1160

17. D. Karatzas, F. Shafait, S. Uchida, M. Iwamura, L.G. i Bigorda, S.R. Mestre, J. Mas, D.F. Mota, J.A. Almazan, L.P. De Las Heras, ICDAR 2013 robust reading competition, in *2013 12th International Conference on Document Analysis and Recognition* (IEEE, 2013), pp. 1484–1493

18. I. Krasin, T. Duerig, N. Alldrin, V. Ferrari, S. Abu-El-Haija, A. Kuznetsova, H. Rom, J. Uijlings, S. Popov, S. Kamali, M. Malloci, J. Pont-Tuset, A. Veit, S. Belongie, V. Gomes, A. Gupta, C. Sun, G. Chechik, D. Cai, Z. Feng, D. Narayanan, K. Murphy, Openimages: a public dataset for large-scale multi-label and multi-class image classification (2017), Dataset available from https://storage.googleapis.com/openimages/web/index.html

19. R. Krishna, Y. Zhu, O. Groth, J. Johnson, K. Hata, J. Kravitz, S. Chen, Y. Kalantidis, L.-J. Li, D.A. Shamma, et al., Visual genome: connecting language and vision using crowdsourced dense image annotations **123**(1), 32–73 (2017)

20. M. Liao, B. Shi, X. Bai, TextBoxes++: a single-shot oriented scene text detector. IEEE Trans. Image Process. **27**(8), 3676–3690 (2018)

21. M. Liao, B. Shi, X. Bai, X. Wang, W. Liu, TextBoxes: a fast text detector with a single deep neural network, in *Thirty-First AAAI Conference on Artificial Intelligence* (2017)

22. X. Liu, D. Liang, S. Yan, D. Chen, Y. Qiao, J. Yan, FOTS: fast oriented text spotting with a unified network, in *Proceedings of the IEEE Conference on Computer Vision and Pattern Recognition* (2018), pp. 5676–5685

23. P. Lyu, M. Liao, C. Yao, W. Wu, X. Bai, Mask textspotter: an end-to-end trainable neural network for spotting text with arbitrary shapes, in *Proceedings of the European Conference on Computer Vision (ECCV)* (2018), pp. 67–83
24. A. Mishra, K. Alahari, C. Jawahar, Image retrieval using textual cues, in *Proceedings of the IEEE International Conference on Computer Vision* (2013), pp. 3040–3047
25. A. Mishra, S. Shekhar, A.K. Singh, A. Chakraborty, OCR-VQA: visual question answering by reading text in images, in *2019 International Conference on Document Analysis and Recognition (ICDAR)* (IEEE, 2019), pp. 947–952
26. J. Pennington, R. Socher, C. Manning, Glove: global vectors for word representation (2014), pp. 1532–1543
27. B. Shi, X. Wang, P. Lyu, C. Yao, X. Bai, Robust scene text recognition with automatic rectification, in *Proceedings of the IEEE Conference on Computer Vision and Pattern Recognition* (2016), pp. 4168–4176
28. A. Singh, V. Goswami, V. Natarajan, Y. Jiang, X. Chen, M. Shah, M. Rohrbach, D. Batra, D. Parikh, MMF: a multimodal framework for vision and language research (2020)
29. A. Singh, V. Natarajan, M. Shah, Y. Jiang, X. Chen, D. Batra, D. Parikh, M. Rohrbach, Towards VQA models that can read (2019)
30. A. Veit, T. Matera, L. Neumann, J. Matas, S. Belongie, COCO-text: dataset and benchmark for text detection and recognition in natural images (2016), arXiv:1601.07140
31. X.-C. Yin, W.-Y. Pei, J. Zhang, H.-W. Hao, Multi-orientation scene text detection with adaptive clustering. IEEE Trans. Pattern Anal. Mach. Intell. **37**(9), 1930–1937 (2015)
32. X. Zhou, C. Yao, H. Wen, Y. Wang, S. Zhou, W. He, J. Liang, EAST: an efficient and accurate scene text detector, in *Proceedings of the IEEE Conference on Computer Vision and Pattern Recognition* (2017), pp. 5551–5560

Chapter 13
Visual Question Generation

Abstract To explore how questions regarding images are posed and abstract the events caused by objects in the image, the visual question generation (VQG) task has been established. In this chapter, we classify VQG methods according to whether their objective is data augmentation or visual understanding.

13.1 Introduction

Natural questions are not focused on what can be seen but rather focused on what can be inferred given the visible objects. To move beyond the literal description of image content and understand the abstract concepts of images, researchers have introduced the novel task of visual question generation (VQG) [11], in which given an image, the system must ask a natural and engaging question. The VQG task can not only be used for data augmentation in the VQA task but also enable the machine to better understand the images. In this chapter, we describe methods for the VQG task. According to the abovementioned concept, we divide the methods based on the purpose of the VQG task: data augmentation or visual understanding. VQG methods aimed at data augmentation are divided into three categories: generating questions from answers (Sect. 13.2.1), generating questions from images (Sect. 13.2.2), and adversarial learning (Sect. 13.2.3).

13.2 VQG as Data Augmentation

The key purpose of VQG as data augmentation is to support the dataset construction for VQA. The existing approaches typically treat the VQG task as a reversed visual question answer (VQA) task, requiring an exhaustive match among all the image regions and the given answer.

13.2.1   Generating Questions from Answers

Motivation

To enhance the robustness of VQA models, researchers have proposed an answer-centric approach to focus on only the relevant image regions. This method can promptly find the core answer area in an image and generate questions that can be addressed by the given answer. This framework ensures that all generated questions can be accurately answered.

Methods

Liu et al. [9] proposed an iVQA model that can gradually adjust its focus of attention guided by both a partially generated question and the answer. This framework is a deep neural network with three subnets: an image encoder (the ResNet-152 model), an answer encoder (LSTM), and a question decoder (LSTM). The two encoders provide inputs for the decoder to generate a sentence that fits the conditioned answer and image content. A multimodal attention module is also a key component that dynamically directs image attention given the outputs of both encoders and a partial question encoder.

Shah et al. [18] proposed a question generation module similar to a conditional image captioning model. The question generation module consists of two linear encoders that transform attended image features obtained from the VQA model and distribution over the answer space to lower dimensional feature vectors. These feature vectors are summed with additive noise and passed through an LSTM that is trained to reconstruct the original question and optimized by minimizing the negative log likelihood with teacher forcing. The module does not pass the one-hot vector representing the answer obtained or an embedding of the answer obtained to the question generation module but rather the predicted distribution over answers. This framework enables the question generation module to learn to map the model's confidence over answers to the generated question.

Liu et al. [10] proposed a variational iVQA model that can generate diverse, grammatically correct and content correlated questions that match a given answer. Specifically, the question encoder encodes the image and questions the mean and variance of a Gaussian distribution. Subsequently, the decoder takes an image feature vector, an answer encoding, and a noise vector as the inputs and generates visual questions. The noise vector is sampled from N (μ,s2•1) and N (0, 1) during training and sampling, respectively.

Xu et al. [21] proposed an approach named the radial graph convolutional network (Radial-GCN), which focuses only on the image regions related to the answers. The Radial-GCN method can promptly find the core answer area in an image by matching the latent answer with the semantic labels learned from all image regions. Subsequently, a novel sparse graph of the radial structure is naturally built to capture the associations between the core node (i.e., answer area) and peripheral nodes (i.e., other areas). Graphic attention is subsequently adopted to steer the convolutional propagation toward potentially more relevant nodes for final question generation.

Performance and Limitations

Modules to generate questions from answers only generate corresponding questions based on answers, ignore global features, and cannot generate more complex and detailed questions or questions based on more features in images, which decreases the diversity of questions and is not suitable in cases involving small datasets.

13.2.2 Generating Questions from Images

Motivation

In contrast to generating questions from answers, modules aimed at generating questions from images can generate varied types of informative questions. This method uses the answers to annotate the images and generates questions from the annotations.

Methods

Kafle et al. [6] proposed two methods for generating QA pairs regarding images: (1) a template-based generation method that uses image annotations and (2) a long short-term memory (LSTM)-based language model. The template data augmentation method uses semantic segmentation annotations to generate new QA pairs. The model synthesizes four kinds of questions from the annotations: yes/no, counting, object recognition, and scene, activity and sport recognition. One key issue with the template-based augmentation method is that the questions are rigid and may not closely resemble the way questions are typically posed in the VQA dataset. To address this aspect, a stacked LSTM that generates questions regarding images is trained. The network consists of two LSTM layers, each with 1,000 hidden units followed by two fully connected layers, with 7,000 units each, corresponding to the size of the vocabulary constructed by tokenizing training questions into individual words. The first fully connected layer has a ReLU activation function, while the second layer has a 7,000-way softmax. The output question is produced one word at a time until the end-of question token is reached.

Ray et al. [14] proposed a question generator that synthesizes questions with similar intent. Specifically, the question generator first concatenates the deep features of an image and concatenates the QA pair to an embedding. Image features are obtained using a ResNet152 framework. QA features are obtained using an embedding layer for each word in the question, which are fed into a 1-layer question-encoder LSTM. The last output of the question-encoder LSTM is concatenated with the deep image features. These concatenated features are fed to another 1-layer LSTM to generate a similar-intent question. The output LSTM is trained using teacher forcing and a cross entropy loss. The top 5 probability-weighted random sampling methods are used in the evaluation process.

Krishna et al. [7] proposed a model that maximizes the mutual information between the image, expected answer and generated question. In the training phase of this model, the image and answer are embedded into a latent space z and reconstructed, thereby maximizing the mutual information with the image and answer. During inference, given an image input and an answer category (e.g., attribute), the model encodes both entities into a latent representation. The model obtains samples from the latent representation with noise to generate questions that are relevant to the image and whose answers result in the given answer category. This framework allows the model to generate goal-driven questions for any image, focused on extracting its objects and attributes, among other aspects.

Sarrouti et al. [16] introduced an approach to generating visual questions regarding radiology images, known as VQGR, i.e., an algorithm that can ask a question when shown an image. VQGR first generates new training data from the existing examples based on contextual word embeddings and image augmentation techniques. Subsequently, the framework uses the variational autoencoder model to encode images into a latent space and decode natural language questions.

Alwatter et al. [1] proposed a deep multilevel attention model to address inverse visual question answering. This model generates regional visual and semantic features at the object level and enhances them with the answer cue by using attention mechanisms. Two levels of multiple attention are employed in the model, including dual attention in the partial question encoding step and dynamic attention in the next question word generation step.

Performance and Limitations

Although the addition of image features to the task of generating questions from images allows for the generation of questions containing richer image information, the main questions are still generated around the answer attention regions with a limited influence of the global (image-wide) features because the annotations are generated based on answers.

13.2.3   Adversarial Learning

Motivation

Question answering (QA) and question generation (QG) have intrinsic connections, and these two tasks can mutually enhance each other. The QA model judges whether the question generated by a QG model is relevant to the answer. In contrast, the QG model provides the probability of generating a question given the answer, which is useful evidence that facilitates QA. Adversarial learning regards QA and QG as dual tasks. The training framework devises an agent for VQG and VQA with pretrained models, and the learning tasks of the two agents form a closed loop, the objectives of which are simultaneously optimized to guide each other via a reinforcement learning process.

Methods

Xu et al. [20] proposed the dual learning framework of a model for VQG. The two agents of VQG and VQA are initially equipped with pretraining models. The learning tasks of the two agents form a closed loop, the objectives of which are simultaneously optimized to guide each other via a reinforcement learning process with specific reward signals as feedback to each agent.

Li et al. [8] proposed an end-to-end unified model, the invertible question answering network (iQAN), which consists of two components for VQA and VQG. The VQA and VQG components are formulated as inverse processes by introducing a novel parameter sharing scheme and the duality regularizer. Input questions and answers are encoded into fixed-length features by using an RNN and a lookup table, respectively. Predicted features are obtained using the attention and MUTAN fusion modules. The predicted features are used to obtain the output (by LSTM and a linear classifier for questions and answers, respectively).

Zhang et al. [22] proposed a deep reinforcement learning framework based on three new intermediate rewards, namely, goal-achievement, progressiveness and informativeness, which encourage the generation of succinct questions. A target object is assigned to Oracle, but it is unknown to the VQG and guesser. Subsequently, the VQG generates a series of questions, which are answered by Oracle. During training, Oracle answers the question based on all the objects in each round and measures the informativeness reward. Moreover, the guesser generates a probability distribution to measure the progressiveness reward. Finally, the number of rounds are considered, and the goal-achievement reward is set based on the success status. These intermediate rewards are adopted for optimizing the VQG agent by the reinforcement.

Fan et al. [3] proposed two discriminators, specifically, natural and human-written discriminators, to enhance the training. The reinforcement learning framework is used to incorporate scores from the two discriminators as the reward to guide the training of the question generator.

Guo et al. [4] proposed a new framework for video question generation, which introduces an attention mechanism to process the inference of the dialog history. A selection mechanism is used to select a question from the candidate questions generated by each round of dialog history. A recent video question answering model is used to predict the answer to the generated question, and the answer quality is used as the reward to fine-tune the model based on a reinforced learning mechanism.

Performance and Limitations

Adversarial learning is performed by continuously feeding new types of adversarial samples for training, thereby continuously enhancing the model robustness. To ensure effectiveness, the method requires the use of high-intensity adversarial samples and a network architecture with sufficient expressiveness. However, imperfections in the training phase of deep neural networks render these frameworks vulnerable to adversarial samples, such as inputs crafted by adversaries with the intent of causing deep neural networks to misclassify.

13.3 VQG as Visual Understanding

In contrast to VQG as data augmentation, VQG as visual understanding no longer relies on the question's answer but generates higher cognitive level questions regarding what can be inferred rather than what can be seen from an image on the basis of scene understanding and prior information regarding the objects.

Motivation

The aim of VQG as visual understanding is to generate questions that have a tightly focused purpose—questions with the aim of learning something specific regarding the image. VQG as visual understanding uses image features as the input to generate questions that have open-ended answers. Since the use of image features alone can lead to excessive attention, image captioning is often introduced to achieve effective alignment between visual and textual representations.

Methods

Jain et al. [5] proposed a creative algorithm for visual question generation that combines the advantages of variational autoencoders with long short-term memory networks. When a variational autoencoder is used, the choice of appropriate LSTM models for the encoder (Q-distribution) and decoder (P-distribution) is of crucial importance. The Q-distribution encodes a given sentence and a given image signal into a latent representation. The V-dimensional 1-hot encoding of the vocabulary is linearly embedded. Embedding and F-dimensional image features are the LSTM inputs, transformed to fit the H-dimensional hidden space. The final hidden representation is transformed via two linear mappings to estimate the mean and log-variance. The P-distribution is used to reconstruct a given question, the image representation, and an M-variate random sample. To obtain a prediction, the H-dimensional latent space is transformed into V-dimensional logits.

Zhang et al. [23] proposed a model that images and captions generated by a dense caption model as the input, samples the most probable question types, and generates the questions in sequence. First, DenseCap is used to construct dense captions that provide an almost complete coverage of information for questions.

Subsequently, these captions are fed into the question-type selector to sample the most likely question types. The question types, dense captions, and visual features generated by VGG-16 are used as the input, and the question generator decodes this information to formulate questions.

Rothe et al. [15] proposed a probabilistic generative model aimed at predicting the questions that people may or may not ask. Parameters of the model are fitted to predict the frequency with which humans ask particular questions in a particular context in the data set by [asking and evaluating natural language questions]. Formally, fitting the generative model is a problem of density estimation in the space of question-like programs, where the space is defined by the grammar.

Patro et al. [12] proposed the use of a multimodal differential network to produce natural and engaging questions. This model contains three main modules: (a) a representation module that extracts multimodal features, (b) a mixture module that fuses the multimodal representation and (c) a decoder that generates questions using an LSTM-based language model.

Fan et al. [2] proposed a question type-driven framework to produce multiple questions for a given image with different foci. In this framework, each question is constructed following the guidance of a sampled question type in a sequence-to-sequence fashion. To diversify the generated questions, a novel conditional variational autoencoder is introduced to generate multiple questions with a specific question type. Moreover, a strategy is formulated to conduct question type distribution learning for each image to select the final questions.

Uehara et al. [19] proposed a method for generating questions regarding unknown objects in an image to obtain information regarding classes that have not been learned. First, objects in the input image are detected using the object region proposal module. Next, the unknown object classification and target selection module identifies whether each object is unknown and selects an object region to be the target of the question. Finally, the visual question generation module generates a question using features extracted from the whole image and target region.

Patro et al. [13] proposed a principled deep Bayesian learning framework that combines multiple visual and language cues to produce natural questions. The model has three experts, namely, place experts, caption experts and tag experts, to provide information (advice) related to different cues. Subsequently, a moderator is used that weighs this advice and passes the resultant embedding to the decoder to generate a natural question.

Scialom et al. [17] proposed BERT-gen, an architecture for text generation based on BERT that can leverage either mono- or multimodal representations. In this work, textual and visual inputs are considered sequences. Captions are encoded via BERT embeddings, while visual embeddings are obtained via a linear layer, used to project image representations to the embedding layer dimensions.

Performance and Limitations

The limitation of this approach is that it has limited effectiveness despite the introduction of image caption tasks for improvement. When generating problems in complex scenarios, challenges associated with generating a single type of question and missing detailed questions are encountered.

References

1. Y. Alwatter, Y. Guo, Inverse visual question answering with multi-level attentions, in *ACML* (2020)
2. Z. Fan, Z. Wei, P. Li, Y. Lan, X. Huang, A question type driven framework to diversify visual question generation, in *IJCAI* (2018)
3. Z. Fan, Z. Wei, S. Wang, Y. Liu, X. Huang, A reinforcement learning framework for natural question generation using bi-discriminators, in *COLING* (2018)
4. Z. Guo, Z. Zhao, W. Jin, Z. Wei, M. Yang, N. Wang, N.J. Yuan, Multi-turn video question generation via reinforced multi-choice attention network. IEEE Trans. Circuits Syst. Video Technol. **31**, 1697–1710 (2021)
5. U. Jain, Z. Zhang, A. Schwing, Creativity: generating diverse questions using variational autoencoders, in *2017 IEEE Conference on Computer Vision and Pattern Recognition (CVPR)* (2017), pp. 5415–5424
6. K. Kafle, M.A. Yousefhussien, C. Kanan, Data augmentation for visual question answering, in *INLG* (2017)
7. R. Krishna, M.S. Bernstein, L. Fei-Fei, Information maximizing visual question generation, in *2019 IEEE/CVF Conference on Computer Vision and Pattern Recognition (CVPR)* (2019), pp. 2008–2018
8. Y. Li, N. Duan, B. Zhou, X. Chu, W. Ouyang, X. Wang, Visual question generation as dual task of visual question answering, in *2018 IEEE/CVF Conference on Computer Vision and Pattern Recognition* (2018), pp. 6116–6124
9. F. Liu, T. Xiang, T.M. Hospedales, W. Yang, C. Sun, iVQA: inverse visual question answering, in *2018 IEEE/CVF Conference on Computer Vision and Pattern Recognition* (2018), pp. 8611–8619
10. F. Liu, T. Xiang, T.M. Hospedales, W. Yang, C. Sun, Inverse visual question answering: a new benchmark and VQA diagnosis tool. IEEE Trans. Pattern Anal. Mach. Intell. **42**, 460–474 (2020)
11. N. Mostafazadeh, I. Misra, J. Devlin, M. Mitchell, X. He, L. Vanderwende, Generating natural questions about an image (2016), arXiv:abs/1603.06059
12. B.N. Patro, S. Kumar, V. Kurmi, V.P. Namboodiri, Multimodal differential network for visual question generation, in *EMNLP* (2018)
13. B.N. Patro, V. Kurmi, S. Kumar, V.P. Namboodiri, Deep Bayesian network for visual question generation, in *2020 IEEE Winter Conference on Applications of Computer Vision (WACV)* (2020), pp. 1555–1565
14. A. Ray, K. Sikka, A. Divakaran, S. Lee, G. Burachas, Sunny and dark outside?! improving answer consistency in VQA through entailed question generation, in *EMNLP/IJCNLP* (2019)
15. A. Rothe, B. Lake, T. Gureckis, Question asking as program generation, in *NIPS* (2017)
16. M. Sarrouti, A.B. Abacha, D. Demner-Fushman, Visual question generation from radiology images, in *ALVR* (2020)
17. T. Scialom, P. Bordes, P.-A. Dray, J. Staiano, P. Gallinari, Bert can see out of the box: on the cross-modal transferability of text representations (2020), arXiv:abs/2002.10832

18. M. Shah, X. Chen, M. Rohrbach, D. Parikh, Cycle-consistency for robust visual question answering, in *2019 IEEE/CVF Conference on Computer Vision and Pattern Recognition (CVPR)* (2019), pp. 6642–6651
19. K. Uehara, A.T. de Pablos, Y. Ushiku, T. Harada, Visual question generation for class acquisition of unknown objects, in *ECCV* (2018)
20. X. Xu, J. Song, H. Lu, L. He, Y. Yang, F. Shen, Dual learning for visual question generation, in *2018 IEEE International Conference on Multimedia and Expo (ICME)* (2018), pp. 1–6
21. X. Xu, T. Wang, Y. Yang, A. Hanjalic, H.T. Shen, Radial graph convolutional network for visual question generation. IEEE Trans. Neural Netw. Learn. Syst. **32**, 1654–1667 (2021)
22. J. Zhang, Q. Wu, C. Shen, J. Zhang, J. Lu, A.V. Hengel, Asking the difficult questions: goal-oriented visual question generation via intermediate rewards, in *ECCV* (2018)
23. S. Zhang, L. Qu, S. You, Z. Yang, J. Zhang, Automatic generation of grounded visual questions (2017), arXiv:abs/1612.06530

Chapter 14
Visual Dialogue

Abstract Visual dialogue is an important and complicated vision language task that processes the visual features of images and textual features of captions, questions and histories to answer questions. To accomplish this task, the machine must exhibit the abilities of perception, multimodal reasoning, relationship mining and visual coreference resolution. In this chapter, we briefly describe the challenges associated with this method and introduce the two benchmarks. Subsequently, a comprehensive review of the associated methods is presented, which are classified into four categories.

14.1 Introduction

Visual dialogue (VD) is a cross-modal task lying at the intersection between computer vision and natural language processing. Given the capacities of reasoning, grounding, recognition and translation, a visual dialog agent is expected to answer questions based on an image, caption and history dialog. Hence, a visual dialog task is related to visual question answering (adding caption and history as the input), visual grounding (converting the visual information located in bounding boxes into human language) and image captioning (generating a description according to the history and question).

As a classic problem in the field of visual language, visual dialogue must simultaneously process inputs from both vision and language modalities. The processing of multimodal inputs can be divided into two parts: perception and reasoning. Perception emphasizes single-modal feature extraction, while reasoning highlights the further interaction and association of multimodal features to obtain a multimodal joint feature representation. Specifically, visual dialogue requires the model to not only understand the intent of the question but also extract the image content corresponding to the question and abstract the historical information related to this question Therefore, the complicated reasoning associated with multimodal features is a considerable challenge for visual dialogue. In addition, in visual dialogue, several pronouns refer to something or someone that appeared previously, which is easy for people but difficult for machines to understand. In particular, the machine must be able to not only resolve the pronouns but also further associate the pronouns

Q. Wu et al., *Visual Question Answering*, Advances in Computer Vision and Pattern Recognition, https://doi.org/10.1007/978-981-19-0964-1_14

with the target objects in the visual scene, which is the visual coreference resolution challenge for visual dialogue. Moreover, visual dialogue encounters a dataset bias problem, mainly a language bias problem. Specifically, in the training phase, the visual dialogue model may rely excessively on the correlation between the question and answer and remember the matching pattern between the question and answer, thereby ignoring the exploration of the image content. Therefore, the performance and robustness are considerably restricted. Solving the language bias and enhancing the versatility and robustness of the model are thus key challenges for the visual dialogue task.

To solve the above mentioned problems, a series of methods have been proposed since the introduction of the visual dialogue task. For vision language reasoning, several attention mechanism-based methods (Sect. 14.3) have been proposed to focus on question-related information, and graph-based methods (Sect. 14.5) have been proposed to mine the relationships among different kinds of features. In addition, a number of visual coreference resolutions (Sect. 14.4) have been proposed to solve the coreference problem. Furthermore, several researchers have introduced pretrained models (Sect. 14.6), which learn visual-semantic knowledge from other vision-language datasets or other tasks to break the dataset bias problem. The experiments are mainly conducted on two benchmarks, i.e., VisDial and GuessWhat?!

In the following section, we describe the datasets and comprehensively review the four categories of methods.

14.2   Datasets

A series of datasets has been established for the task of visual dialogue. In this section, we describe the existing two mainstream visual dialogue datasets, along with their construction mechanism and their main characteristics (Table 14.1).

VisDial

VisDial [1], one of the benchmarks for visual dialogue, is available in two versions, v0.9 and v1.0. VisDial-v0.9 is collected through a game, the context of which is based on the images and captions collected from the MSCOCO dataset [7]. For a conversation regarding an image, two annotators implement the annotation through

Table 14.1   Major datasets for visual dialog and their main characteristics

Dataset	Number of images	Number of QA pairs	Number of dialogues	Source of images
VisDial [1]	133,351	1,261,510	133,351	MSCOCO
GuessWhat?! [2]	66,537	821,889	160,745	MSCOCO and Flickr

an interactive game. In the game, one annotator plays the role of the questioner, and the other annotator plays the role of the answerer. The questioner can only see the caption and the conversation history, not the image, while the answerer can see the caption, conversation history and image. To understand the image content, the questioner asks successive questions regarding the invisible image. The respondent provides an answer based on the questioner's questions, combining images and the conversation history. Through this data collection process, each image is matched with 10 rounds of question and answer conversations. VisDial-v0.9 is divided into two subsets: a training set and a validation set. The collection process of VisDial-v1.0 is the same as that of VisDial-v0.9. VisDial-v1.0 is divided into three subsets: the training set, validation set, and test set. The training set of VisDial-v1.0 is composed of all the data from VisDial-v0.9, and the images and dialogues are obtained based on the MSCOCO dataset. The validation and test sets for VisDial-v1.0 are based on Flickr images [16]. The validation set of VisDial-v1.0 contains 2,000 dialogues, and the test set contains 8,000 dialogues.

GuessWhat?!

GuessWhat?! [2] is a large-scale dataset consisting of 150 K human-played games with a total of 800 K visual question-answer pairs on 66 K images. This dataset pertains to a cooperative two-player game in which both players see the picture of a rich visual scene with several objects. One player–the oracle–is randomly assigned an object (which could be a person) in the scene. This object is not known by the other player–the questioner–whose goal is to locate the hidden object. To this end, the questioner can ask a series of yes-no questions that are answered by the oracle.

14.3 Attention Mechanism

A general neural network recognizes objects by training a neural network with a large amount of data. For example, a neural network that has been trained over a large number of handwritten digits can recognize the value represented by a new handwritten digit. However, a neural network trained in this way is actually equivalent to processing the full features of a picture. Although the neural network learns the features of the image for classification, these features are not different in the "eyes" of the neural network, and the neural network does not focus excessively on a particular "region". In general, when humans look at a picture, they focus their attention on a region of the picture. In addition to grasping a picture as a whole, humans focus on the local information of the picture, such as the location of a local table and types of goods, while other information receives less attention. The basic idea of the attention mechanism in computer vision is to ensure that computers learn to ignore irrelevant information and focus on the key information. In the visual dialogue task, the approach based on the attention mechanism accurately captures the subject information of questions and images by weighting the attention to the question or image and enhancing the interaction between vision and language. Based on the understanding of the proposed problem and the historical dialogue, importance weights are assigned to regions of the image to identify the most relevant region to the problem. In this section, we introduce several typical approaches based on the attention mechanism.

14.3.1 Hierarchical Recurrent Encoder with Attention (HREA) and Memory Network (MN)

Das et al. [1] introduced two baseline models with attention mechanism, i.e., the hierarchical recurrent encoder with attention and the memory network. The HREA extracts the features of the conversation history after extracting the features of the image and question and subsequently performs the attention-weighted calculation for each word in the conversation history. The MN performs the attention-weighted calculation for each conversation history based on the features of the image and question.

Hierarchical Recurrent Encoder with Attention (HREA). The HREA considers only the last question-answer pair in the past as the conversation history, and after extracting the features of images and questions, computes the attention weights for each word in the conversation history to extract the features of the conversation history. As shown in Fig. 14.1, the HREA involves a dialog-RNN sitting atop a recurrent block (R_t). The recurrent block R_t embeds the question and image jointly via an LSTM, embeds each round of the history H_t, and passes a concatenation of these entities to the dialog-RNN above it. The dialog-RNN produces both an encoding for this round (E_t in Fig. 14.1) and a dialog context to pass onto the next round. Moreover, there exists an attention-over-history mechanism that allows the recurrent block R_t to choose and attend to the round of the history relevant to the current question. This attention mechanism consists of a softmax over previous rounds $(0, 1, \ldots, t - 1)$ computed from the history and question+image encoding.

Memory Network (MN). The MN computes attention weights for each conversation history based on the features of the image and the question, stores each question-answer pair as a "fact" and answers the current question based on the facts.

However, all of these methods directly utilize the sentence feature for the dialog history and question while using the flattened feature for the image, considering only the overall information of the sentence and image at a high level and ignoring the detailed information of words in the sentence and regions in the image at a low level.

14.3.2 History-Conditioned Image Attentive Encoder (HCIAE)

Motivation. A common approach is to use an encoder architecture with an attention mechanism that implicitly performs coreference resolution by identifying the portion

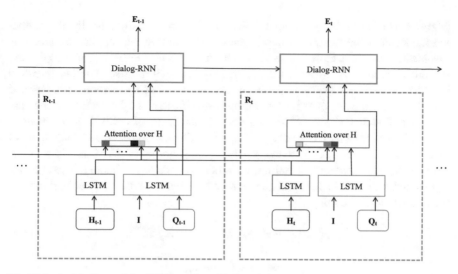

Fig. 14.1 Architecture of the HRE encoder with attention

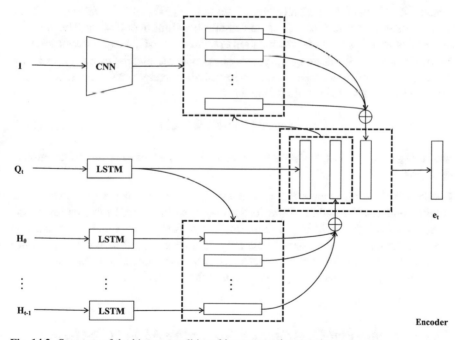

Fig. 14.2 Structure of the history-conditioned image attentive encoder

of the dialog history that can help in answering the current question while considering a holistic representation for the image. Intuitively, one would expect that the answer is localized to regions in the image and is consistent with the attended history. With this motivation, Lu et al. [8] proposed the HCIAE, as shown in Fig. 14.2.

Method. The HCIAE uses the spatial feature of the image. Specifically, this framework considers the sequential dependency of the dialog and applies an attention mechanism to select the relevant information of the dialog history to supplement the information of the question. Subsequently, the method uses another attention mechanism to select relevant spatial regions of the image to capture the targeted visual information for question answering. Specifically, the HCIAE uses the spatial image features $V \in \mathcal{R}^{d \times k}$ from a convolution layer of a CNN. q_t is encoded with an LSTM to obtain a vector $m_t^q \in \mathcal{R}^d$. Simultaneously, each previous round of history $(H_0, \ldots, H_{t-1})$ is encoded separately with another LSTM as $M_t^h \in \mathcal{R}^{d \times t}$. Conditioned on question embedding, the model attends to the history. The attended representation of the history and question embedding are concatenated and used as input to attend to the image:

$$z_t^h = w_a^T \tanh(W_h M_t^h + (W_q m_t^q) \Bbbk^T) \tag{14.1}$$

$$\alpha_t^h = \text{softmax}(z_t^h) \tag{14.2}$$

where $\Bbbk \in \mathcal{R}^t$ is a vector with all elements set as 1. $W_h, W_q \in \mathcal{R}^{t \times d}$ and $w_a \in \mathcal{R}^k$ are parameters to be learned. $\alpha \in \mathcal{R}^k$ is the attention weight over history. The attended history feature $\hat{m}_t^h$ is a convex combination of columns of M_t, weighted appropriately by the elements of α_t^h. Subsequently, we concatenate m_t^q and $\hat{m}_t^h$ as the query vector and obtain the attended image feature $\hat{v}_t$ in a similar manner. All three components are used to obtain the final embedding e_t:

$$e_t = \tanh(W_e[m_t^q, \hat{m}_t^h, \hat{v}_t]) \tag{14.3}$$

where $W_e \in \mathcal{R}^{d \times 3d}$ is the weight parameter, and $[,]$ represents the concatenation operation.

Limitation. The history-conditioned image attentive encoder only considers the detailed region information of the image and uses the overall information of the sentence for the dialog history and question. The framework ignores the fact that words in the sentence also contain detailed information regarding the dialog history and question. In addition, this approach directly uses the spatial feature of the image while ignoring the region-to-region relation in the image.

14.3.3 Sequential Co-Attention Generative Model (CoAtt)

Motivation. The existing visual dialog systems oversimplify the training objectives and focus only on measuring the word-level correctness. Moreover, the produced responses tend to be generic and repetitive. For example, a simple response of 'yes', 'no', or 'I don't know' can safely answer a large number of questions and lead to a high MLE objective value. The generation of more comprehensive answers and

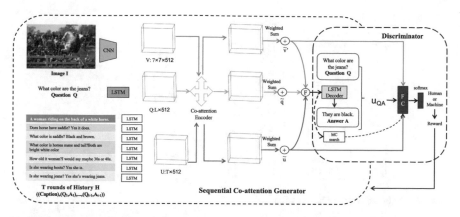

Fig. 14.3 Adversarial learning framework with a sequential co-attention generative model. The model is composed of two components: a sequential co-attention generator that accepts the image, question and dialog tuples as the input and uses the co-attention encoder to jointly analyze them; and a discriminator that identifies whether each answer is generated by a human or the generative model by considering the attention weights. The output from the discriminator is used as a reward to push the generator to generate responses that are indistinguishable from those that a human might generate

a deeper engagement of the agent in the dialog requires a more engaged training process. A satisfactory dialog generation model should generate responses indistinguishable from those produced by humans. In contrast to VQA, which has only one round of questioning, visual dialogues have multiple rounds of conversation, and the conversation history needs to be accessed and understood. In this context, it is desirable to establish an encoder that can combine multiple sources of information. A naive approach is to separately represent the input images, histories, and questions and connect them to learn a joint representation. However, it is more powerful to ensure that the model selectively focuses on regions of the image and segments of the conversation history based on the question. Considering these aspects, Wu et al. [15] proposed an adversarial learning-based approach to generate visual dialog.

Method. As shown in Fig. 14.3, the traditional dialog generator is used; i.e., the image, question and dialog history are encoded using CNN and LSTM, and a weight is assigned to each local representation by using the co-attention model. Subsequently, the local feature is summed with the weights to obtain an attended feature, and the feature is decoded using the LSTM to obtain a corresponding answer. The key point in this model is that a discriminator is added to the back of the model to distinguish whether the input answer is human- or machine-generated. The input is not only the corresponding question and answer but also the output of the attention to ensure that the discriminator can analyze whether the question and answer are reasonable under certain circumstances. The probabilities generated by the discriminator are used as a reward for the generator to update the parameters of the generator.

The attention model in the generator is a sequential co-attention model, as shown in Fig. 14.4. Specifically, the framework refers to an encoder-decoder style generative

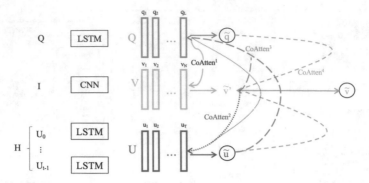

Fig. 14.4 Sequential co-attention encoder

model that has been widely used in sequence generation problems. The model first uses a pretrained CNN [13] to extract the spatial image features $V = [v_1, \ldots, v_N]$ from the convolutional layer, where N is the number of image regions. The question features are $Q = [q_1, \ldots, q_L]$, where $q_l = LSTM(w_l, q_{l-1})$, which is the hidden state of an LSTM at step l given the input word w_l of the question. L is the length of the question. Because the history H is composed of a sequence of utterances, the model extracts each utterance feature separately to identify the dialog history features; i.e., $U = [u_0, \ldots, u_T]$, where T is the number of rounds of the utterance (QA-pairs). Each u is the last hidden state of an LSTM, which accepts the utterance word sequences as the input. Given the encoded image, dialog history and question features V, U and Q, the proposed model uses a co-attention mechanism to generate attention weights for each feature type using the other two as the guidance in a sequential style. Each co-attention operation is denoted as $\tilde{x} = \text{CoAtten}(X, g_1, g_2)$, which can be expressed as follows:

$$H_i = \tanh(W_x x_i + W_{g_1} g_1 + W_{g_2} g_2) \tag{14.4}$$

$$\alpha_i = \text{softmax}(W^T H_i), \quad i = 1, \ldots, M, \tag{14.5}$$

$$\tilde{x} = \sum_{i=1}^{M} \alpha_i x_i \tag{14.6}$$

where X is the input feature sequence (i.e., V, U or Q), and $g_1, g_2 \in \mathbb{R}^d$ represent the guidance as the outputs of previous attention modules. d is the feature dimension. W_x, W_{g_1}, $W_{g_2} \in \mathbb{R}^{h \times d}$ and $W \in \mathbb{R}^h$ are learnable parameters. h denotes the size of the hidden layers of the attention module. M is the input sequence length that corresponds to N, L and T for different feature inputs.

As shown in Fig. 14.4, in the proposed process, the initial question feature is used to attend to the image. The weighted image features and initial question representation are combined to attend to the utterances in the dialog history to produce the attended dialog history ($\tilde{u}$). The attended dialog history and weighted image region features are jointly used to guide the question attention ($\tilde{q}$). Finally, the image attention ($\tilde{v}$) is implemented, guided by the attended question and dialog history, to complete the circle. All three co-attended features are concatenated and embedded to the final feature F:

$$F = \tanh(W_{eg}[\tilde{v}; \tilde{u}; \tilde{q}]) \tag{14.7}$$

where [;] represents a concatenation operator. Finally, this vector representation is fed to an LSTM to compute the probability of generating each token in the target using a softmax function, which forms the response $\hat{A}$. The generation process is represented as $\pi(\hat{A} \mid V, U, Q)$.

Limitation. Although the proposed sequential co-attention model leverages co-attention mechanisms to capture cross-modal correlations, its reasoning ability is limited. The model typically concatenates the multimodal features and directly projects the concatenated feature into the answer feature space by a neural network. The reasoning process does not fully utilize the rich relational information in this task due to the monolithic vector representations of dialog. Moreover, the associated feedforward network fails to deeply and iteratively mine and reason the information from different dialog entities over the inherent dialog structures.

14.3.4 Synergistic Network

Motivation. Classical visual dialog systems integrate the image, question, and history to search for or generate the best matched answer, and thus, this approach ignores the role of the answer. Considering this aspect, Guo et al. [3] proposed a novel image-question-answer synergistic network to value the role of the answer for precise visual dialog.

Method. The synergistic network shown in Fig. 14.5 extends the traditional one-stage solution to a two-stage solution. In the first stage, known as the primary stage, representative vectors of the image, dialog history, and initial question are learned.

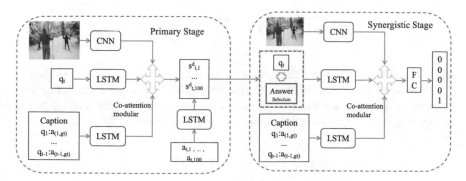

Fig. 14.5 Architecture of the synergistic network. All candidate answers are scored in the primary stage, and certain selected answers are rescored in the synergistic stage

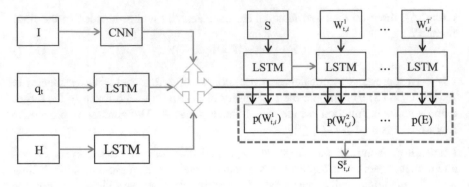

Fig. 14.6 Primary stage of synergistic network. The score of each answer is the associated probability of the word sequence

To this end, objects and their features are detected in the input image using Faster-RCNN. These features are encoded using a convolutional neural network. As the question and dialog history contain text data, these entities are encoded using the LSTM. All the candidate answers that are generated in the primary stage are scored based on their relevance to the image and question pair. In the second stage, the synergistic stage is dubbed, and the answers in synergy with the image and question are ranked based on their probability of correctness.

An encoder-decoder solution is adopted in the primary stage, as shown in Fig. 14.6. The encoder performs two main tasks: dereferencing in multiple turn conversations and locating the objects in the image mentioned in the current question. The attention mechanism is commonly used to perform tasks. Instead of linear concatenation, this framework chooses multimodal factorized bilinear pooling (MFB) [17], as this framework can overcome the difference between the distributions of the two features (two LSTMs to encode the question and history; LSTMs for the text feature; and CNNs for the image feature). In the MFB, the fusion of two features, X and Y, $\in R^d$ is calculated as follows:

$$z = \mathrm{MFB}(X, Y) = \sum_{i=1}^{k}(\mathbf{U}_i^\top X \circ \mathbf{V}_i^\top Y) \qquad (14.8)$$

where $\mathbf{U}$ and $\mathbf{V} \in R^{d \times l \times k}$ are the parameters to be learned, k is the number of factors, l is the hidden size, and $\circ$ is the Hadamard product (elementwise multiplication). However, Y sometimes represents multiple channel inputs, e.g, detected objects or history in this model, and thus, the formula can be expressed as follows:

$$z = \mathrm{MFB}(X, Y) = \sum_{i=1}^{k}((\mathbf{U}_i^\top X \cdot \mathbb{K}^\top) \circ (\mathbf{V}_i^\top Y)) \qquad (14.9)$$

where $\not{\mathbb{K}} \in R^{\phi}$ is the vector with all elements equal to one, and ϕ is the channel number of Y.

Limitation. *The two-stage method is more time intensive than the one-stage method.*

14.4 Visual Coreference Resolution

In natural language, people often use pronouns and abbreviations to refer to the same word to avoid repetition of words. Pronouns lead to a lack of clarity and incomplete structure and limit the understanding of natural language by machines. Due to the presence of references, the respondent needs to not only disambiguate the reference but also associate the reference with the target objects in the visual scene to accurately understand the questioner's intent and answer. Therefore, visual pronoun disambiguation is proposed for machines to realize visual dialogues and complex visual reasoning. This process of visual reference resolution is key to accurately localize attention in the presence of ambiguous expressions and is thus of significance in extending VQA approaches to the visual dialogue task. To solve the visual reference resolution problem, many approaches have been proposed in recent years.

Methods

AMEM Seo et al. [12] proposed a visual dialogue model with attention memory-based reference resolution. This framework uses a memory network to memorize and store each visual attention computed from the historical question and answer pairs and weighs the stored visual attention according to the current question to perform visual referential disambiguation at the sentence level. The framework retrieves the previous visual attention maps by applying soft attention over all the memory dictionaries and concatenating it with the current visual attention.

CorefNMN Kottur et al. [6] proposed a neural module network architecture for visual dialog. CorefNMN combines symbolic computation and neural networks to decompose the visual inference process into several basic operations and stores the entities that appear in the conversation history through a referent pool. When a referent is encountered, CorefNMN correlates the referent with the target visual object through the query module, thereby achieving visual referential disambiguation at the word level.

RvA Niu et al. [9] proposed the recursive visual attention approach, which adopts a recursive strategy. Specifically, the approach first determines whether the current question is clear before answering it. If the question is unclear, the question is back-tracked to the question that best matches the topic of the current question, and the process is recursively repeated until the question is clear and the recursion is terminated. Through the process of recursive backtracking, the RvA approach explicitly implements visual denotational disambiguation at the word level.

DAN DAN [5] consists of two kinds of attention modules, REFER and FIND. Specifically, the REFER module learns latent relationships between a given question and a dialog history by employing a multihead attention mechanism. The FIND module considers the image features and reference-aware representations (i.e., output of the REFER module) as the input and performs visual grounding via a bottom-up attention mechanism.

Limitation

AMEM and CorefNMN use only word-level or sentence-level representations and encounter limitations in identifying the semantic intent of the question. Both approaches and RvA also involve limitations in that they store all previous visual attention, while studies on the human memory system show that visual sensory memory, due to its rapid decay property, does not store all previous visual attention.

14.5 Graph-Based Methods

With the advancement of the extant research, a series of methods have been developed that are not limited to learning entity representations but can also mine the relationships among entities. Since graphs have the natural character of representing entities and their relationships, there are several methods for visual dialogue that employ graph networks to represent the feature embeddings of images and dialogues and the relations among them. According to the categories of the entities contained in the graph network, the graph structure utilized by these methods can be divided into two types: single-modality and cross-modality graph structures.

14.5.1 Scene Graph for Visual Representations

Motivation

Visual dialogue involves multiple questions that cover a broad range of visual content that can be related to objects, relationships or semantics. Jiang et al. [4] indicated that the existing models simply extract visual features as monolithic representations and thus have a limited expressive ability when addressing variant questions. Therefore, the authors attempted to adaptively capture question-relevant fine-grained visual information by employing scene graphs to abstract object embeddings and their relationships.

Method

Since it is necessary to consider the objects in an image and their relationships when addressing complex questions, DualVD approach simultaneously uses the object

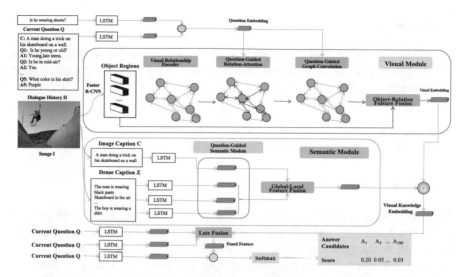

Fig. 14.7 Overview of the DualVD model for visual dialogue. The model contains two parts: the visual module and semantic module, where "G" represents the gate operation given a question and history dialogue. The visual module is constructed using a scene graph

embeddings and the relationships between objects within a scene graph, as shown in Fig. 14.7. In the scene graph, the objects and relationships between these objects are represented by nodes and edges, respectively. Furthermore, a pretrained Faster-RCNN [10] and a pretrained visual relationship encoder [18] are employed to extract the initial embeddings for the objects and relationships between them. After the initialization, the scene graph is utilized to capture question-relevant object embeddings and relationship embeddings to accurately reply to the current question. In practice, the current question feature is fused with the history dialogue to generate a comprehensive history-aware question embedding, which is utilized to guide the capture of question-relevant object instances and relationships. Thereafter, question-guided relation attention is adopted to learn a series of critical scores for all the object relationships according to their correlations to the generated question embedding. The object relationships are weighted by the corresponding attention scores. Subsequently, the object features, i.e., the nodes in the scene graph, are refined by a question-guided graph convolution module. During the refinement stage, for each node, the relationship features with the neighbors of the nodes and the features of their neighbors are concatenated and used to calculate a relevant score with question embedding. The obtained relevant score is regarded as the adjacency between the current node and its neighbor. After this refinement, all the refined relation-aware object features are fused with the original object features via an object-relation information fusion module to ensure the appropriate proportion of object appearance and the visual relationships contained in the updated object representations. Finally, the visual feature of the whole image is calculated by fusing all the obtained updated object representations.

Limitation

The above mentioned kind of graph network constructs the scene graph only for visual embeddings and does not mine the fine-grained information of the historical dialogue and question.

14.5.2 GNN for Visual and Dialogue Representations

Motivation

The underlying semantic dependencies between dialog entities are essential for visual dialogue, while the existing methods largely neglect the rich relation information in the dialog. Although a few methods leverage co-attention mechanism to capture the cross-modal correlations, they fail to deeply and iteratively mine and reason the information from different dialog entities, and thus, the reasoning ability is limited. To address this problem, Zheng et al. [19] and Schwartz et al. [11] proposed the construction of a GNN to represent the visual dialogue, in which the nodes denote dialog entities and the edges indicate semantic dependencies between dialog entities, which enables deep and iterative mining and reasoning.

Method

VisDial-GNN. The architecture of VisDial-GNN for visual dialogue is exhibited in Fig. 14.8. The graph is constructed with the observed Q&A nodes, unobserved answer nodes and their relations as edges. First, the embedding for each node is initialized by fusing the image feature and language embedding of the corresponding sentence(s) via a co-attention layer, as shown in the feature embedding module in Fig. 14.8. After initializing the node hidden states with feature embeddings, the iterative inference is initiated by the expectation-maximization (EM) algorithm, which involves an M-step

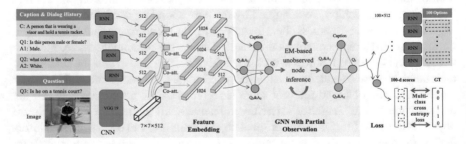

Fig. 14.8 The visual dialog is represented by a GNN, in which the nodes represent the dialog entities (i.e., caption, question and answer pairs, and the unobserved queried answer), and the edges denote the semantic dependencies between nodes

(estimating the edge weights) and an E-step (updating the embedding for unobserved nodes). Subsequently, the hidden state of the unobserved node is regarded as the answer embedding, which is fused with the predefined answer candidates to compute the loss. A multiclass cross-entropy loss on the fused embeddings is used to train the graph neural network.

Factor Graph Attention. The algorithm of factor graph attention is exhibited in Fig. 14.9. The factor graph is defined over utilities, which, in the visual dialog setting, include an image I, an answer A, a caption C, and a history of past interactions $(H_{Q_t}, H_{A_t})_{t\in\{1,...,T\}}$. Each utility consists of basic entities; e.g, a question is composed of a sequence of words, and an image is composed of spatially ordered regions.

First, the image utility and textual utilities are initialized by the embeddings from a pretrained CNN model and LSTM model. Subsequently, the representation of each utility is updated by the two types of factors in the factor graph, as illustrated in Fig. 14.10. Local factors capture information within a utility, such as their entity representation and local interactions, while joint factors capture interactions of any subset of utilities. The utility representation is updated by the attention mechanism, in which the attention value is obtained under the guidance of local and joint factors. Finally, the algorithm fuses the utility representations with each predefined answer candidate and produces a posterior probability for each answer. The model is trained using the maximum likelihood method.

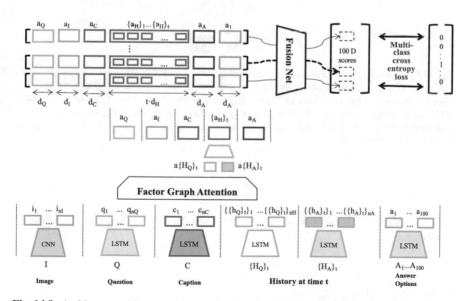

Fig. 14.9 Architecture of factor graph attention for visual dialogue

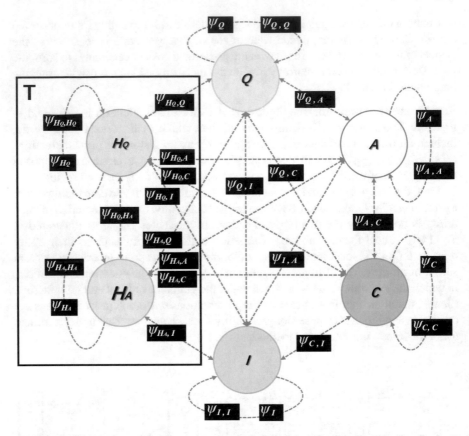

Fig. 14.10 Graphical representation of the attention-based factor graph. Specifically, Q, A, C, I, H_A and H_Q denote the question, answer, caption, image, history answer and history question, respectively. There exist two kinds of factors: (1) local factors that capture information within a utility, such as their entity representation, i.e., ψ_Q and their local interactions, i.e., $\psi_{Q,Q}$, and (2) joint factors that capture the interactions of any subset of utilities, i.e., $\psi_{Q,A}$. T is the number of history dialog interactions

Limitation

This kind of graph-based method only supports answer ranking for predefined answer candidates.

14.6   Pretrained Models

Pretrained models have been noted to be effective in addressing various vision-language tasks. In this section, we introduce several representative visual dialogue algorithms employing pretraining models, which are based on the transformer architecture.

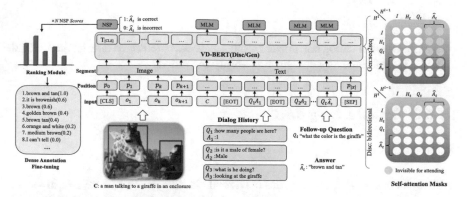

Fig. 14.11 Model architecture of VD-BERT for both discriminative and generative settings

14.6.1 VD_BERT

Motivation

The existing methods mostly focused on unibidirectional attention mechanisms to model the implicated interactions, i.e., to learn the attention from answer to question, image regions or dialog history. However, these methods cannot comprehensively and mutually consider the relationship between all the entities (image, question, history dialogue and answer), thereby failing to exploit the provided multimodality information. To fully capture the intricate interactions between all the entities, Wang et al. [14] proposed a unified visual-dialog architecture that simultaneously receives all the entities as input and can capture all the mutual relations between these entities via a transformer-based bidirectional attention mechanism. The authors pretrained the employed transformer model on a vision-language dataset to ensure that it can manage multimodality inputs.

Method

The architecture is exhibited in Fig. 14.11. First, a unified vision-dialog transformer is used to capture the relations between all the entities, in which a pretrained BERT is employed to preliminarily initialize the designed model. As the employed BERT is specifically pretrained to address the language input, two visually grounded training tasks, i.e., masked language modeling (MLM) and next sentence prediction (NSP), are adopted to pretrain the VD-BERT on the VisDial dataset [1], thereby allowing the model to simultaneously manage multimodality inputs. Thereafter, this fully pretrained model is fine-tuned in a vision-dialog task with a ranking optimization module.

Limitation

The unified vision-dialog transformer architecture was pretrained only on the VisDial dataset [1], and thus, its generalization ability is limited.

14.6.2   Visual-Dialog BERT

Motivation

Although considerable progress has been made in the visual dialogue task in recent years, most of this progress occurred in isolation, and deep neural networks were trained only on the VisDial dataset. These methods ignored the significant amount of shared knowledge in related vision-language tasks (e.g, captioning and visual question answering) that can benefit visual dialog frameworks. Therefore, Jiang et al. [4] pretrained their model on other related vision and language datasets and transferred the knowledge to visual dialog to boost the performance of visual dialog.

Method

To operate the two types of information, i.e., image and text, the authors adapt the recently proposed ViLBERT, which has two transformer-based encoders, one encoder for each of the two modalities, i.e., language and vision. The interaction between the two modalities is enabled by co-attention layers.

The training process is shown in Fig. 14.12. First, the language stream is pretrained on English Wikipedia and BooksCorpus [20] datasets with the masked language modeling (MLM) and next sentence prediction (NSP) tasks. Next, to learn powerful visually grounded representations before fine-tuning on the VisDial dataset, the model is trained on the large-scale conceptual captions and visual question answering datasets with simple yet powerful self-supervised tasks, i.e., masked image region (MIR), MLM and NSP. Finally, the model is fine-tuned on sparse annotations from VisDial [1] with the MIR, MLM and NSP losses and optionally fine-tuned on dense annotations.

Limitation

The model only supports answer ranking and does not support answer generation.

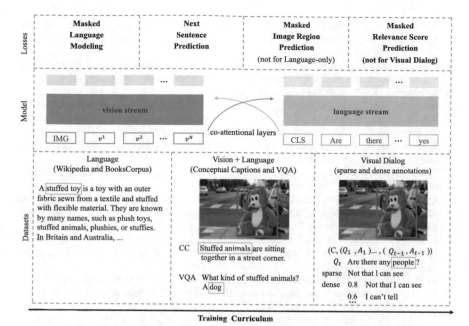

Fig. 14.12 Visual-dialog BERT

References

1. A. Das, S. Kottur, K. Gupta, A. Singh, D. Yadav, J.M. Moura, D. Parikh, D. Batra, Visual dialog, in *Proceedings of the IEEE Conference on Computer Vision and Pattern Recognition* (2017), pp. 326–335
2. H. De Vries, F. Strub, S. Chandar, O. Pietquin, H. Larochelle, A. Courville, Guesswhat?! visual object discovery through multi-modal dialogue, in *Proceedings of the IEEE Conference on Computer Vision and Pattern Recognition* (2017), pp. 5503–5512
3. D. Guo, C. Xu, D. Tao, Image-question-answer synergistic network for visual dialog, in *Proceedings of the IEEE/CVF Conference on Computer Vision and Pattern Recognition* (2019), pp. 10434–10443
4. X. Jiang, J. Yu, Z. Qin, Y. Zhuang, X. Zhang, Y. Hu, Q. Wu, Dualvd: an adaptive dual encoding model for deep visual understanding in visual dialogue, in *Proceedings of the AAAI Conference on Artificial Intelligence*, vol. 34 (2020), pp. 11125–11132
5. G.-C. Kang, J. Lim, B.-T. Zhang, Dual attention networks for visual reference resolution in visual dialog, in *Proceedings of the 2019 Conference on Empirical Methods in Natural Language Processing and the 9th International Joint Conference on Natural Language Processing (EMNLP-IJCNLP)* (2019), pp. 2024–2033
6. S. Kottur, J.M. Moura, D. Parikh, D. Batra, M. Rohrbach, Visual coreference resolution in visual dialog using neural module networks, in *Proceedings of the European Conference on Computer Vision (ECCV)* (2018), pp. 153–169
7. T.-Y. Lin, M. Maire, S. Belongie, J. Hays, P. Perona, D. Ramanan, P. Dollár, C.L. Zitnick, Microsoft coco: common objects in context, in *ECCV* (Springer, 2014), pp. 740–755
8. J. Lu, A. Kannan, J. Yang, D. Parikh, D. Batra, Best of both worlds: transferring knowledge from discriminative learning to a generative visual dialog model

9. Y. Niu, H. Zhang, M. Zhang, J. Zhang, Z. Lu, J.-R. Wen, Recursive visual attention in visual dialog, in *Proceedings of the IEEE/CVF Conference on Computer Vision and Pattern Recognition* (2019), pp. 6679–6688

10. S. Ren, K. He, R. Girshick, J. Sun, Faster R-CNN: towards real-time object detection with region proposal networks. IEEE Trans. Pattern Anal. Mach. Intell. **39**(6), 1137–1149 (2016)

11. I. Schwartz, S. Yu, T. Hazan, A.G. Schwing, Factor graph attention, in *Proceedings of the IEEE/CVF Conference on Computer Vision and Pattern Recognition* (2019), pp. 2039–2048

12. P.H. Seo, A. Lehrmann, B. Han, L. Sigal, Visual reference resolution using attention memory for visual dialog, in *Advances in Neural Information Processing Systems* (2017)

13. K. Simonyan, A. Zisserman, Very deep convolutional networks for large-scale image recognition (2014), arXiv:1409.1556

14. Y. Wang, S. Joty, M.R. Lyu, I. King, C. Xiong, S.C. Hoi, Vd-bert: a unified vision and dialog transformer with bert (2020), arXiv:2004.13278

15. Q. Wu, P. Wang, C. Shen, I. Reid, A. Van Den Hengel, Are you talking to me? reasoned visual dialog generation through adversarial learning, in *Proceedings of the IEEE Conference on Computer Vision and Pattern Recognition* (2018), pp. 6106–6115

16. P. Young, A. Lai, M. Hodosh, J. Hockenmaier, From image descriptions to visual denotations: new similarity metrics for semantic inference over event descriptions. Trans. Assoc. Comput. Linguist. **2**, 67–78 (2014)

17. Z. Yu, J. Yu, J. Fan, D. Tao, Multi-modal factorized bilinear pooling with co-attention learning for visual question answering, in *Proceedings of the IEEE International Conference on Computer Vision* (2017), pp. 1821–1830

18. J. Zhang, Y. Kalantidis, M. Rohrbach, M. Paluri, A. Elgammal, M. Elhoseiny, Large-scale visual relationship understanding, in *Proceedings of the AAAI Conference on Artificial Intelligence*, vol. 33 (2019), pp. 9185–9194

19. Z. Zheng, W. Wang, S. Qi, S.-C. Zhu, Reasoning visual dialogs with structural and partial observations, in *Proceedings of the IEEE/CVF Conference on Computer Vision and Pattern Recognition* (2019), pp. 6669–6678

20. Y. Zhu, R. Kiros, R. Zemel, R. Salakhutdinov, R. Urtasun, A. Torralba, S. Fidler, Aligning books and movies: towards story-like visual explanations by watching movies and reading books, in *Proceedings of the IEEE International Conference on Computer Vision* (2015), pp. 19–27

Chapter 15
Referring Expression Comprehension

Abstract Referring expression comprehension (REC) aims to localize objects in images based on natural language queries. In contrast to the object detection task, in which queried object labels are predefined, the REC problem can only observe the queries during the test. REC is difficult to implement because this task requires a comprehensive understanding of complicated natural language and various types of visual information. In this chapter, we first describe this task and subsequently introduce prevalent datasets proposed for REC tasks such as the RefCOCO, RefCOCO+ and RefCOCOg datasets. Finally, we classify the methods in the REC domain into three main categories: two-stage models, one-stage models and reasoning process comprehension.

15.1 Introduction

Referring expression comprehension (REC) aims to localize objects in images based on natural language queries. This aspect is an essential block in the field of human–machine interaction, and it can also facilitate other downstream tasks such as vision-language navigation [4], image retrieval [1] and visual dialogue [26]. While significant progress has been made in computer vision and natural language processing, REC remains challenging because this task not only needs to address various types of visual information but also requires a comprehensive understanding of attributes, relationships and contextual information by language. More importantly, unlike object detection, the REC system must use language to select the best objects from many candidates rather than using predefined category labels to classify these regions. Therefore, many studies [12, 13, 18, 21, 25, 27] have attempted to better solve this problem from different perspectives.

In this chapter, we review the referring expression comprehension (REC) from two aspects: datasets and models. The dataset introduces four kinds of mainstream datasets: ReferItGame, RefCOCO, RefCOCO+, RefCOCOg and Flickr30k entities. Subsequently, we comprehensively review the methods to implement REC. These methods can be divided into two categories: two-stage models (Sect. 15.3) and one-stage models (Sect. 15.4). The two-stage models can be divided into three categories: joint embedding, co-attention models and graph-based models.

15.2 Datasets

A number of datasets have been proposed for the referring expression comprehension task. In the following sections, we introduce the existing mainstream referring expression datasets, in terms of the dataset construction and main characteristics of the datasets. The dataset characteristics are summarized in Table 15.1.

ReferItGame

ReferItGame [6] is the first large-scale referring expression dataset for real-world scenes, which contains natural images from the ImageCLEF IAPR [5] dataset with segmented regions from the SAIAPR-12 [3] dataset. The dataset is collected by a two-player interactive game, in which the first player generates expressions referring to objects in images, and the second player needs to click the correct locations according to the descriptions of objects. Based on this game, the ReferItGame dataset has produced 130,525 expressions referring to 96,654 distinct objects in 19,894 images. However, this dataset focuses mostly on context rather than objects, and images often only have one object for a given class, which allows for speaker models to generate short descriptions without taking into context the ambiguity.

RefCOCO and RefCOCO+

RefCOCO [24] and RefCOCO+ [24] have also been collected in the ReferItGame [6] scenario, in which players try to generate efficient information to indicate the correct objects to the other player. In RefCOCO, the type of language used in the referring expressions is not subject to any limitations, whereas RefCOCO+ disallows the use of location words and focuses on purely appearance-based descriptions. The images of these datasets are obtained from the MSCOCO [8] dataset. RefCOCO contains 142,209 referring expressions to 50,000 objects in 19,994 images, and RefCOCO+ generates 141,564 expressions for 49,856 objects in 19,992 images. The datasets are split into training, validation, testA and testB. The testA and testB sets contain only people and nonpeople, respectively.

Table 15.1 Major datasets for referring expression comprehension and their main characteristics

Dataset	Number of images	Number of expressions	Number of objects	Avg. length words	Source of images
ReferItGame [6]	19,894	130,525	96,654	3.61	Image CLEF
RefCOCO [24]	19,994	142,209	50,000	3.61	MSCOCO
RefCOCO+ [24]	19,992	141,564	49,856	3.53	MSCOCO
RefCOCOg [13]	26,711	104,560	54,822	8.43	MSCOCO
Flickr30k Entities [15]	31,783	158,915	275,775	–	Flickr30k

RefCOCOg

RefCOCOg [13] was collected in noninteractive scenarios on Amazon Mechanical Turk. A group of workers were required to write natural language expressions of the object in the images of MSCOCO, and another group of workers was requested to click on the specified object given the referring expression. If the click area overlapped with the correct object, then the referring expression was valid and added to the data set. If not, then another expression was considered for the object. RefCOCOg introduced 85,474 referring expressions for 54,822 objects in 26,711 images. The average lengths of referring expressions in RefCOCO and RefCOCO+ are 3.61 and 3.65, respectively, while in RefCOCOg, the average length is 8.43 words.

Flickr30k entities

The Flickr30k entities [15] consist of 31,783 images, which expands the 158k captions from the Flickr30k dataset [22] with 224k coreference chains and contains 276k bounding box annotations. The annotation process is divided into two stages: forming coreference chains that refer to the same entities and annotating bounding boxes for the resulting chains. This workflow can reduce redundancy by identifying coreferent mentions, and second, coreference annotation is intrinsically valuable, e.g., for training cross-caption coreference models.

15.3 Two-Stage Models

In this section, we introduce two-stage methods for referring expression comprehension. These methods are mainly composed of two stages. The first step is to generate candidate regions using pretrained detectors such as Faster-RCNN [16]. In the second stage, each region is compared to the input query, and a similarity score is output. During inference, the region with the highest similarity score is output as the final prediction. In the two-stage framework, various studies differ from one another in terms of the second step based on insights. We describe the two-stage methodology in three subsections focused on joint embedding methods, co-attention methods and graph-based methods. The basic idea of each method is introduced in the following text.

15.3.1 Joint Embedding

Motivation. The main concept of joint embedding is to learn a mapping relation between vision and language by embedding them into the same feature space. Specifically, as shown in Fig. 15.1 for the representation of images, these methods usually use convolutional neural networks (CNNs) to generate rich image representations and

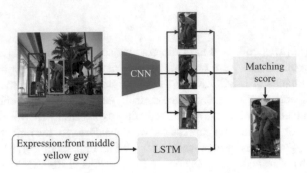

Fig. 15.1 Illustration of the joint embedding method. These methods usually use CNN and LSTM networks to extract image representations and encode language information, respectively. Subsequently, vision and language are embedded in the common space, and the matching score between regions and textual phrases is calculated to select the best match objects

embed input images into fixed-length vectors. Since language represents sequential structure data, a long short-term memory (LSTM) network is used, which encodes the entire language as a single embedding vector. After embedding the visual and textual representations into a common space, a distance metric is learned, and the referred regions are ranked by calculating the distance similarity conditioned on the embedded expression vectors.

Methods. Mao et al. [13] developed the first model to refer to expression comprehension by deep learning. This approach utilizes a convolutional neural network to extract visual representations from proposals and uses an LSTM model to encode the language features. As a baseline model, this method uses a ranking-based approach to select the best region. The authors generate a set of proposals, and the model ranks these regions by probability. This strategy is similar to image retrieval [7, 14], with the only difference being that the images are replaced by regions. Furthermore, this method can solve the problem of referring expression generation. In this subtask, the maximum mutual information method is used to generate the reference expression, which can generate sentences that distinguish the input proposals from other candidates.

Yu et al. [24] attempted to simultaneously address the problems of comprehension and generation. In contrast to the work of Mao et al. [13], the authors focus on encoding comparisons with the most relevant objects instead of using a general feature to encode the context overall images. The authors introduce a visual appearance difference representation, which can represent the difference between the target region and surrounding objects. In particular, the authors select five comparison regions of the same category as the context information to enhance the performance. In addition, the authors use the coordinates of regions to encode the relative location and size differences between the target object and other surrounding objects. With this context modeling, this model exhibits a high performance for referring expression generation and comprehension.

Zhang et al. [25] extended the existing methods to learn context information via variational Bayesian analysis, which can utilize the relationship between the referent and surrounding information. The authors considered that either target or context can influence the estimation of the posterior distribution, and the search space of the context can be reduced by the proposed approach. Specifically, the model consists of three modules: context posterior, referent posterior and context prior. For each object, the model first computes a coarse context, which can help refine the target object of the referring expression. Subsequently, each module aligns the image features with cue-specific textual features to help localize the objects. In this manner, the framework reduces the complexity of the context and achieves a high performance.

Limitations. The joint embedding framework is effective and simple. However, these approaches are limited by the use of global vector representations that ignore complex language semantics and various types of visual information. Therefore, when the models address complex images (such as images containing multiple similar objects) or long sentences, it is difficult to attend to significant image regions and language words.

15.3.2 Co-Attention Models

Motivation. Attention mechanisms have been applied to many deep-learning frameworks. This framework ensures that models focus on an important part of the input when processing high-dimensional features or redundant information. The use of co-attention has been proposed in VQA [11], as a variant of attention. Co-attention highlights the areas that the model must seek and words that it must read in the language. Through its introduction in the referring expression comprehension field, the co-attention mechanism has achieved several achievements. Specifically, the model can build fine-grit connections between visual and textual information to ensure that the system can utilize the features from several regions of interest (ROIs) when encoding each word in the text, and vice versa, leading to semantically enriched visual and textual representations.

Methods. Zhuang et al. [27] considered that conventional frameworks embed visual and language features into a joint space for one-step reasoning; however, when the expression is long or complex, such a one-step process cannot relate multiple parts of the expression to the image. To solve this problem, a parallel attention framework has been established to recurrently attend to objects. This framework includes two parallel attention mechanisms: image attention and region attention. The image attention module encodes the entire image and referring text by recurrently attending to different image regions. This module allows the model to learn helpful context information. In contrast, the region attention module recurrently attends to the candidates that are conditioned by the referring descriptions. Finally, the matching module utilizes the image-level and region-level representations as the input to compute a matching probability for each proposal.

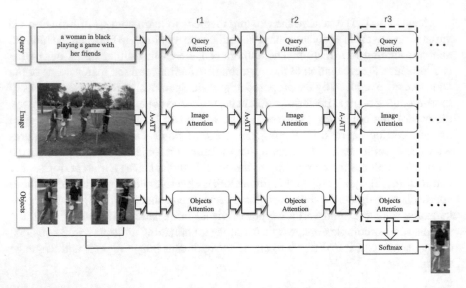

Fig. 15.2 Illustration of the accumulated attention method. Blue, green and yellow represent the attention for the query, image, and objects, respectively. A-ATT conducts more rounds to refine the attention for information communication and accumulation

Deng et al. [2] formulated the referring expression comprehension task into three sequential subtasks: (1) refine the main concept in the language, (2) understand the focus in the image, and (3) search the most relevant region. The authors proposed an accumulated attention method (A-ATT) to simultaneously solve the above mentioned three problems. As shown in Fig. 15.2, this framework adopts three modules to extract the query, images and object attention. The method employs an accumulating process to jointly integrate three types of attention and solve them in a circular manner to capture the correlations among those subtasks. In this way, each type of attention module can be utilized as guidance when computing the other two aspects. Finally, the accumulated attention computes a similarity score between the attended features of each proposal. The refined representations from the language and image attention modules are used to select the target regions.

Co-attention methods are usually combined with other frameworks (such as graph-based models and one-stage models), as introduced in the following sections.

Limitations. The co-attention mechanism can attend to image regions and textual information; however, these attention-based methods cannot guarantee a correct attention assignment since the datasets usually provide no corresponding annotations. Furthermore, these approaches do not consider the complexity of the relationship between multiple regions.

15.3.3 Graph-Based Models

Motivation. The key to solving the referring expression task is to learn the distinguishing object features that can adapt to the expression. To avoid ambiguity, this expression usually describes not only the properties of the object itself but also the relationship between the object and its neighbors. The existing methods manage only the objects or study only first-order relationships between objects without considering the potential complexity of expressions. Therefore, graph-based methods have been proposed, in which the nodes can highlight related objects, and the edges are used to recognize the object relationship existing in the expression.

Methods. Wang et al. [18] proposed a language-guided graph attention method (LGRAN). As shown in Fig. 15.3, LGRANs includes three modules, namely, the language-self attention module, language-guided graph attention module, and matching module. The first module utilizes a self-attention mechanism to decompose the language of three parts (relationships, intraclass relationships and interclass relationships). The language-guided graph attention constructs candidate objects of a directed graph, intraclass edges and interclass edges. Finally, each region obtains three types of expression-relevant representations. The matching module computes the similarity score for each object. Moreover, the LGRANs can dynamically enrich the region representations based on the attended graph to suit the language. In addition, LGRANs visualize the attention distributions over objects and relationships, which provides an effective basis for understanding the reasoning interpretability of the method.

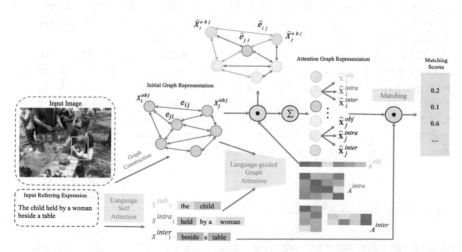

Fig. 15.3 Overview of the language-guided graph attention network. This model is composed of three modules: (1) the language self-attention module parses the expression into subjects, intraclass relationships and interclass relationships, (2) the language-guided graph attention module establishes a directed graph between objects, and (3) the matching module, which computes the matching scores between expressions and objects

To explore the potential complex relationships between objects and expressions, Yang et al. [19] proposed a dynamic graph attention network (DGA), which can achieve multistep reasoning of the interactions between the image and language. Given an image and language, this model builds a graph of objects in the image, in which the nodes are objects and edges are relationships. Similarly, the words of language are also integrated into the graph. Next, the differential analyzer models the expression guidance of reasoning by exploring the structure of the language and updates the compound object representation at every node. Through the guidance of the predicted visual reasoning process, this model performs dynamic inference in a sequential manner on the graph. Finally, this model calculates the similarity score between the compound region and referring expression. The DGA method achieves high performance through this multistep reasoning on top of the relationships between objects in images.

Liu et al. [10] considered that the global context and its interrelationships of grounding objects are important to achieve the correct reasoning. The authors built a new language-guided graph to capture the global context of grounding entities and their relations and learn context-aware cross-modal matching for the visual grounding task. This framework is divided into four parts: (1) an encoder network extracts the features of the language and image, (2) a phrase graph network improves the initial word embedding features by adding phrase relation clues in the description, (3) a visual graph network enriches the features of objects and their context through message propagation over the visual object graph, and (4) the referring expression task, which is considered a graph matching problem between the phrase and visual object graph. A graph-based similarity network is introduced to predict the node and edge similarity scores between the language and regions. In this manner, the method can perform global matching between the visual and textual graph nodes and relation edges to learn the cross-modal context for visual grounding.

Limitations. Although graph-based models can effectively manage the relationship between the multiple related objects conditioning on language, they encounter two problems: (1) bridging the gap between the unstructured data to structured data (for instance, the word order is destroyed when we use the graph to model the language information) and (2) learning the other types of features (such as color, size and post) in addition to the relationship. Therefore, there remains considerable scope for improvement in the area of graph-based models.

15.4   One-Stage Models

Motivation. Although the existing two-stage methods exhibit a satisfactory performance, these frameworks are capped by the first stage with inevitable error accumulation (if the target object cannot be captured in the first stage, the frameworks fail regardless of the performance of the second-ranking stage). Moreover, an existing study [21] demonstrated that two-stage methods incur high computational costs. For each object proposal, both feature extraction and cross-modality similarity com-

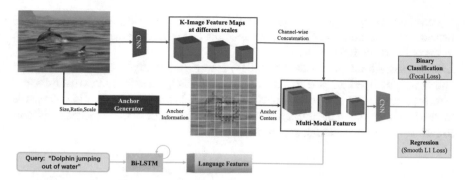

Fig. 15.4 Overview of the ZSGNet architecture. An image-query pair is input to the network. The image is used to produce feature maps of different scales, and the query is encoded to language features through LSTM. The anchor generator produces anchors at different scales and resolutions. Subsequently, the multimodal feature maps that contain images, anchors and language are input to the full convolution network to predict anchor matching scores

putation must be conducted. In contrast to the two-stage approaches, one-stage approaches can directly predict bounding boxes and exhibit a high speed and accuracy.

Methods. Yang et al. [21] proposed a one-stage referring expression comprehension model based on YOLOv3. In particular, a feature pyramid is extracted from the image using Darknet, and the language features are obtained from the referring expression by BERT. Subsequently, each level feature map and the entire language feature are concatenated and fed into the detection head to produce the bounding box of the target. Furthermore, to ensure that the model has the sensory ability of the location, spatial features that consist of normalized coordinates also concatenate the feature map to input the detection head.

Sadhu et al. [17] developed a zero-shot grounding method (ZSGNet) that integrates the detection framework and referring expression comprehension system. Given an image with fixed candidate boxes, the comprehension task is to select the best candidate box (also known as the anchor box) and regress it to a tight bounding box. As shown in Fig. 15.4, the authors adopt a ResNet with a feature pyramid to extract different scales of image feature maps and utilize Bi-LSTM to extract the language representation. In addition, this model also utilizes the focal loss to classify the regions inside/outside the target objects. In this manner, ZAGNet directly learns to locate objects in an end-to-end manner and is significantly superior to the existing system in the zero-shot setting.

Luo et al. [12] proposed a multitask collaborative network that jointly solves referring expression comprehension and referring expression segmentation in a one-stage framework. Specifically, the authors attempt to solve the conflict problem between the two referring expression tasks by (1) consistency energy maximization, which compels the two tasks to assign similar attention to the input image, and (2) adaptive soft nonlocated suppression (ASNLS), which suppresses the response of unrelated regions in referring expression segmentation based on referring expression comprehension. Furthermore, ASNLS allows the model to have a higher error tolerance in terms of the detection results.

Limitations. Although one-stage methods have a significantly higher inference speed and exhibit competitive performance, these methods ignore the contextual information, especially in the case of complex expressions. Moreover, the existing one-stage methods cannot obtain the inference process from the model, which makes it difficult to realize follow-up research and improvement.

15.5   Reasoning Process Comprehension

Motivation. In the task of referring expression comprehension, the existing studies mainly focused on how to achieve superior fusion between the cross-model information. These methods ignore the interpretability of the reasoning process for the entire vision-language system. Therefore, several visual grounding methods have been established to address this problem, and the main concept of these studies is to link images with the parsed language to obtain a comprehensive understanding of the referring. In this manner, the process of entire reasoning is captured, which can facilitate follow-up work and corresponding technical innovation.

Methods. Yu et al. [23] proposed the modular attention network (MAttNet). First, this method divides the description into three components: subject, location and relationship. Subsequently, the authors design three corresponding vision modules. The subject module deals with category, colors, size and other attributes. The location module manages the absolute and relative locations. The relationship module focuses on contextual relationships. Each module processes a different structure and learns the parameters within its own modular space without affecting the others. In addition, instead of relying on an external language parser, MAttNet learns to parse language automatically through an attention mechanism. Finally, the matching scores of the three visual modules are calculated to measure the similarity between the image and language. Owing to the use of the modular network, the entire reasoning process can be easily obtained.

Liu et al. [9] proposed a cross-modal attention-guided erasing strategy for a comprehension system. This method adopts MAttNet [23] as the backbone. By erasing the most attended part from either language or image information, this strategy impels the model to discover more latent cues for reasoning. Moreover, several modifications are introduced to the original modular network. Specifically, in contrast to [23] the model using only the cross-modal information to learn word-level and module-level attention, this method considers the global features of image and language. Moreover, the method formulates the location and relationship modules into a unified structure with sentence-level attention.

Yang et al. [20] proposed a recursive subquery construction framework that recursively alleviates the referring ambiguity with differently constructed subqueries. As shown in Fig. 15.5, this model represents the intermediate understanding of the refer-

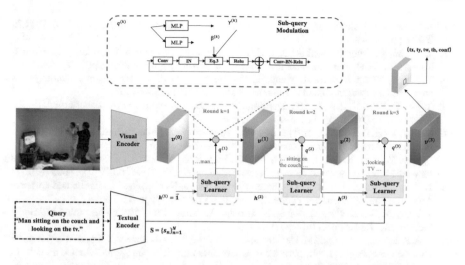

Fig. 15.5 Architecture for the recursive subquery construction framework. This method performs multiple rounds of reasoning on images and expressions by constructing a subquery learner and a modulation network to gradually decrease the ambiguity of referring expression

ring in each round as the text-conditional visual feature, which starts as the image feature and is updated after multiple rounds, terminating as the fused visual-text feature ready for box prediction. In each round, the model constructs a new subquery as a group of words attended with scores to refine the visual feature. Such a multiround solution is in contrast to existing one-stage approaches, as this method can understand the operations of the model and explain its success or failure.

References

1. K. Chen, T. Bui, C. Fang, Z. Wang, R. Nevatia, AMC: attention guided multi-modal correlation learning for image search, in *Proceedings of the IEEE Conference on Computer Vision and Pattern Recognition* (2017), pp. 2644–2652
2. C. Deng, Q. Wu, Q. Wu, F. Hu, F. Lyu, M. Tan, Visual grounding via accumulated attention, in *CVPR* (2018), pp. 7746–7755
3. H.J. Escalante, C.A. Hernández, J.A. Gonzalez, A. López-López, M. Montes, E.F. Morales, L.E. Sucar, L. Villasenor, M. Grubinger, The segmented and annotated IAPR TC-12 benchmark. Comput. Vis. Image Underst. **114**(4), 419–428 (2010)
4. C. Gan, Y. Li, H. Li, C. Sun, B. Gong, VQS: linking segmentations to questions and answers for supervised attention in VQA and question-focused semantic segmentation, in *Proceedings of the IEEE International Conference on Computer Vision* (2017), pp. 1811–1820
5. M. Grubinger, P. Clough, H. Müller, T. Deselaers, The IAPR TC-12 benchmark: a new evaluation resource for visual information systems, in *International Workshop OntoImage*, vol. 2 (2006)
6. S. Kazemzadeh, V. Ordonez, M. Matten, T. Berg, Referitgame: referring to objects in photographs of natural scenes, in *EMNLP* (2014), pp. 787–798

7. R. Kiros, R. Salakhutdinov, R. Zemel, Multimodal neural language models, in *ICML* (PMLR, 2014), pp. 595–603

8. T.-Y. Lin, M. Maire, S. Belongie, J. Hays, P. Perona, D. Ramanan, P. Dollár, C.L. Zitnick, Microsoft COCO: common objects in context, in *ECCV* (Springer, 2014), pp. 740–755

9. X. Liu, Z. Wang, J. Shao, X. Wang, H. Li, Improving referring expression grounding with cross-modal attention-guided erasing, in *CVPR* (2019), pp. 1950–1959

10. Y. Liu, B. Wan, X. Zhu, X. He, Learning cross-modal context graph for visual grounding, in *AAAI*, vol. 34 (2020), pp. 11645–11652

11. J. Lu, J. Yang, D. Batra, D. Parikh, Hierarchical question-image co-attention for visual question answering. Adv. Neural Inf. Process. Syst. **29**, 289–297 (2016)

12. G. Luo, Y. Zhou, X. Sun, L. Cao, C. Wu, C. Deng, R. Ji, Multi-task collaborative network for joint referring expression comprehension and segmentation, in *CVPR* (2020), pp. 10034–10043

13. J. Mao, J. Huang, A. Toshev, O. Camburu, A.L. Yuille, K. Murphy, Generation and comprehension of unambiguous object descriptions, in *CVPR* (2016), pp. 11–20

14. J. Mao, W. Xu, Y. Yang, J. Wang, Z. Huang, A. Yuille, Deep captioning with multimodal recurrent neural networks (m-RNN) (2015)

15. B.A. Plummer, L. Wang, C.M. Cervantes, J.C. Caicedo, J. Hockenmaier, S. Lazebnik, Flickr30k entities: collecting region-to-phrase correspondences for richer image-to-sentence models, in *ICCV* (2015), pp. 2641–2649

16. S. Ren, K. He, R. Girshick, J. Sun, Faster R-CNN: towards real-time object detection with region proposal networks. Adv. Neural Inf. Process. Syst. **28**, 91–99 (2015)

17. A. Sadhu, K. Chen, R. Nevatia, Zero-shot grounding of objects from natural language queries, in *ICCV* (2019), pp. 4694–4703

18. P. Wang, Q. Wu, J. Cao, C. Shen, L. Gao, A.V.D. Hengel, Neighbourhood watch: referring expression comprehension via language-guided graph attention networks, in *CVPR* (2019), pp. 1960–1968

19. S. Yang, G. Li, Y. Yu, Dynamic graph attention for referring expression comprehension, in *ICCV* (2019), pp. 4644–4653

20. Z. Yang, T. Chen, L. Wang, J. Luo, Improving one-stage visual grounding by recursive sub-query construction, in *ECCV* (Springer, 2020), pp. 387–404

21. Z. Yang, B. Gong, L. Wang, W. Huang, D. Yu, J. Luo, A fast and accurate one-stage approach to visual grounding, in *ICCV* (2019), pp. 4683–4693

22. P. Young, A. Lai, M. Hodosh, J. Hockenmaier, From image descriptions to visual denotations: new similarity metrics for semantic inference over event descriptions. Trans. Assoc. Comput. Linguist. **2**, 67–78 (2014)

23. L. Yu, Z. Lin, X. Shen, J. Yang, X. Lu, M. Bansal, T.L. Berg, MAttNet: modular attention network for referring expression comprehension, in *CVPR* (2018), pp. 1307–1315

24. L. Yu, P. Poirson, S. Yang, A.C. Berg, T.L. Berg, Modeling context in referring expressions, in *ECCV* (Springer, 2016), pp. 69–85

25. H. Zhang, Y. Niu, S.-F. Chang, Grounding referring expressions in images by variational context, in *CVPR* (2018), pp. 4158–4166

26. Z. Zheng, W. Wang, S. Qi, S.-C. Zhu, Reasoning visual dialogs with structural and partial observations, In *Proceedings of the IEEE/CVF Conference on Computer Vision and Pattern Recognition* (2019), pp. 6669–6678

27. B. Zhuang, Q. Wu, C. Shen, I. Reid, A. Van Den Hengel, Parallel attention: a unified framework for visual object discovery through dialogs and queries, in *CVPR* (2018), pp. 4252–4261

Part V
Summary and Outlook

This section represents the last part of this book. We summarize the content of this book and highlight possible directions for future research in the domain of visual question answering.

Chapter 16
Summary and Outlook

16.1 Summary

Visual question answering is a significant topic in current AI research and has been linked to many applications such as AI assistant and dialog systems. As a cross-disciplinary task, this topic has attracted considerable attention from researchers in different communities, such as computer vision and natural language processing. VQA is a typical cross-modal task since it requires machines to simultaneously understand visual content (images and videos) and natural language, and, in certain cases, common sense knowledge. Nevertheless, certain challenges must be addressed to realize artificial general intelligence.

In this book, we first present preliminary knowledge regarding deep learning and the task of question-and-answering to provide proper context for the readers. Subsequently, we describe several typical methods for image-based VQA and video-based VQA. For the former, we first describe the classical VQA problem and classical solutions and then focus on the most recent technologies, knowledge base VQA and pretraining for VQA. For the latter, we mainly focus on video presentation learning and several heuristic models. Finally, we describe several advanced topics in VQA, such as embodied VQA, VQA in the form of dialogue, and referring expression comprehension, to extend the readers' horizon.

16.2 Future Directions

16.2.1 Explainable VQA

Most existing VQA models work as a black box: It is not clear how and why these models make predictions or the factors that their decisions are based on. The performance enhancement of such black box methods tends to level off before the expected value. To promote development in this field, it is urgent and necessary to clarify why and how these models work.

The attention mechanism is promising in this direction; however, this mechanism only visualizes an attention map (like a heat map) on the image (or question) to highlight the parts that are important to answer that question. No clear reasoning chain exists to show how and why the model obtains an answer.

The VQA machine [17] represents one further step. The attention mechanism is used but not directly applied to the images. Instead, this mechanism is applied to a collection of supporting facts that are extracted from the image. The attended supporting facts are translated to human readable sentences as the reasons for answering that question. However, the supporting facts are fragmented.

A trusted and explainable VQA system must be able to collect relevant information and correlate them to answer questions and provide trusted explanations. To this end, the machine must fully understand and chain images, questions and knowledge and perform reasoning on the chain.

16.2.2   Bias Elimination

Bias exists in not only collected datasets but also real-world scenarios. For example, we see more red apples than green apples, and we more frequently ride a bike than wheel a bike. Therefore, when a green apple or a person wheeling a bike is present in a picture, most existing models may answer red or riding if we query the color of the apples being eating or the activity being performed by the person. To eliminate this "bias", two possible solutions are as follows: the inclusion of approximately the same amount of data across all scenarios and the enhancement of the reasoning ability of models (i.e., enabling the models to become aware of why they are formulating a certain prediction).

Several recent studies [2, 8, 13] demonstrated that many VQA models answer questions without reasoning and relying excessively on superficial correlations (i.e bias) between the question and answers. To alleviate the challenges associated with VQA, many existing methods mainly focus on weakening the language bias [3, 12].

For example, many methods [3–7, 9–13, 15, 16, 18, 19] have been proposed to overcome the language bias in VQA. These methods can be categorized into two classes, specifically, those with and without data augmentation. Augmentation-based methods [1, 4, 6, 10, 14, 16, 19] seek to balance the biased dataset for unbiased training, while non-augmentation-based methods [3, 5, 7, 9, 11–13, 18] seek to reduce the language bias explicitly or improve attention on the image.

In terms of non-augmentation-based methods, Ramakrishnan et al. [12] adopted adversarial learning between the VQA model and question-only model to prevent the VQA model from capturing the language bias. Inspired by [12], Cadene et al. [3] dynamically adjusted the weight of the samples based on how biased the samples were. In addition, several methods introduced human-based visual [13] and text [18] explanations to strengthen the visual grounding. However, these methods require human annotations that are difficult to obtain. Furthermore, Niu et al. [11] introduced the cause effect to examine the language bias and proposed a counterfactual inference framework to reduce the bias. However, this framework led to the introduction of additional parameters in the inference.

In the context of augmentation-based methods, to ensure that the VQA model focuses on critical objects and words, Chen et al. [4] proposed a CSS method to produce massive counterfactual samples by masking critical objects and words and assigning the corresponding ground-truth answers. To fully exploit the samples, Liang et al. [10] modeled the relationships among original, factual and counterfactual samples to promote the learning of high-level features. In addition, Mutant [6] generated the samples by semantic transformations of the original images or questions. Moreover, without introducing additional annotations, several methods [14, 19] built negative samples from the available samples to balance the dataset.

Furthermore, another issue is that certain "bias" captured from the dataset may represent the natural rule in the real world, i.e common-sense knowledge. Such bias is not harmful to the models, and in fact, the models may benefit from it. For example, "dog" is a kind of "animal", and the color of "oranges" is normally "orange". Thus, filtering and removing the true negative biases in language and vision modalities remain a challenging task.

16.2.3 Additional Settings and Applications

The existing VQA tasks capture only a part of real-world scenarios. Tons of thousands of areas remain untouched, such as VQA for education scenarios and VQA for driving/flying/diving. Challenges for models usually vary as scenario changes. Future work can focus on applying the existing VQA technologies to more applications to facilitate our lives.

VQA has been introduced in many domains, such as medical VQA, to address the questions raised by medical practitioners and patients. Moreover, this concept has been applied to robotics, for example, in the form of embodied VQA, which enables a virtual robot to answer questions in a simulated environment. We believe that VQA can be integrated into more applications under different settings.

References

1. E. Abbasnejad, D. Teney, A. Parvaneh, J. Shi, A. van den Hengel, Counterfactual vision and language learning, in *Proceedings of the IEEE Conference on Computer Vision and Pattern Recognition* (2020), pp. 10041–10051
2. A. Agrawal, D. Batra, D. Parikh, A. Kembhavi, Don't just assume; look and answer: overcoming priors for visual question answering, in *Proceedings of the IEEE Conference on Computer Vision and Pattern Recognition* (2018), pp. 4971–4980

3. R. Cadène, C. Dancette, H. Ben-younes, M. Cord, D. Parikh, RUBi: reducing unimodal biases for visual question answering (2019), pp. 839–850

4. L. Chen, X. Yan, J. Xiao, H. Zhang, S. Pu, Y. Zhuang, Counterfactual samples synthesizing for robust visual question answering, in *Proceedings of the IEEE Conference on Computer Vision and Pattern Recognition* (2020), pp. 10797–10806

5. I. Gat, I. Schwartz, A.G. Schwing, T. Hazan, Removing bias in multi-modal classifiers: regularization by maximizing functional entropies (2020)

6. T. Gokhale, P. Banerjee, C. Baral, Y. Yang, MUTANT: a training paradigm for out-of-distribution generalization in visual question answering, in *Proceedings of the Conference on Empirical Methods in Natural Language Processing* (2020), pp. 878–892

7. C. Jing, Y. Wu, X. Zhang, Y. Jia, Q. Wu, Overcoming language priors in VQA via decomposed linguistic representations, in *Proceedings of the Conference on AAAI* (2020), pp. 11181–11188

8. K. Kafle, C. Kanan, An analysis of visual question answering algorithms, in *Proceedings of the IEEE International Conference on Computer Vision* (2017), pp. 1983–1991

9. G. KV, A. Mittal, Reducing language biases in visual question answering with visually-grounded question encoder, in *Proceedings of the European Conference on Computer Vision* (2020), pp. 18–34

10. Z. Liang, W. Jiang, H. Hu, J. Zhu, Learning to contrast the counterfactual samples for robust visual question answering, in *Proceedings of the Conference on Empirical Methods in Natural Language Processing* (2020), pp. 3285–3292

11. Y. Niu, K. Tang, H. Zhang, Z. Lu, X. Hua, J. Wen, Counterfactual VQA: a cause-effect look at language bias (2020), arXiv:2006.04315

12. S. Ramakrishnan, A. Agrawal, S. Lee, Overcoming language priors in visual question answering with adversarial regularization (2018), pp. 1548–1558

13. R.R. Selvaraju, S. Lee, Y. Shen, H. Jin, S. Ghosh, L.P. Heck, D. Batra, D. Parikh, Taking a HINT: leveraging explanations to make vision and language models more grounded, in *Proceedings of the IEEE International Conference on Computer Vision* (2019), pp. 2591–2600

14. D. Teney, E. Abbasnejad, K. Kafle, R. Shrestha, C. Kanan, A. van den Hengel, On the value of out-of-distribution testing: an example of Goodhart's law (2020)

15. D. Teney, E. Abbasnejad, A. van den Hengel, Learning what makes a difference from counterfactual examples and gradient supervision, in *Proceedings of the European Conference on Computer Vision* (2020), pp. 580–599

16. D. Teney, E. Abbasnejad, A. van den Hengel, Unshuffling data for improved generalization (2020), arXiv:2002.11894

17. P. Wang, Q. Wu, C. Shen, A. van den Hengel, The VQA-machine: learning how to use existing vision algorithms to answer new questions, in *Proceedings of the IEEE Conference on Computer Vision and Pattern Recognition* (2017), pp. 1173–1182

18. J. Wu, R.J. Mooney, Self-critical reasoning for robust visual question answering (2019), pp. 8601–8611

19. X. Zhu, Z. Mao, C. Liu, P. Zhang, B. Wang, Y. Zhang, Overcoming language priors with self-supervised learning for visual question answering, in *Proceedings of the International Joint Conference on Artificial Intelligence* (2020), pp. 1083–1089

Index

Printed in the United States
by Baker & Taylor Publisher Services